GUIDED NOTEBOOK FOR MYMATHLAB®

DEVELOPMENTAL MATHEMATICS

Kirk Trigsted
University of Idaho

Kevin Bodden
Lewis & Clark Community College

Randy Gallaher
Lewis & Clark Community College

PEARSON

Boston Columbus Indianapolis New York San Francisco Upper Saddle River
Amsterdam Cape Town Dubai London Madrid Milan Munich Paris Montreal Toronto
Delhi Mexico City São Paulo Sydney Hong Kong Seoul Singapore Taipei Tokyo

The author and publisher of this book have used their best efforts in preparing this book. These efforts include the development, research, and testing of the theories and programs to determine their effectiveness. The author and publisher make no warranty of any kind, expressed or implied, with regard to these programs or the documentation contained in this book. The author and publisher shall not be liable in any event for incidental or consequential damages in connection with, or arising out of, the furnishing, performance, or use of these programs.

Reproduced by Pearson from electronic files supplied by the author.

Copyright © 2014 Pearson Education, Inc.
Publishing as Pearson, 75 Arlington Street, Boston, MA 02116.

All rights reserved. No part of this publication may be reproduced, stored in a retrieval system, or transmitted, in any form or by any means, electronic, mechanical, photocopying, recording, or otherwise, without the prior written permission of the publisher. Printed in the United States of America.

ISBN-13: 978-0-321-88022-2
ISBN-10: 0-321-88022-6

5 6 7 V069 17 16 15 14

www.pearsonhighered.com

PEARSON

Table of Contents

Module 1 Whole Numbers
1.1 Study Tips for This Course 1
1.2 Introduction to Whole Numbers 5
1.3 Adding and Subtracting Whole Numbers; Perimeter 9
1.4 Multiplying Whole Numbers; Area 13
1.5 Dividing Whole Numbers 17
1.6 Exponents and Order of Operations 21
1.7 Introduction to Variables, Algebraic Expressions, and Equations 25

Module 2 Integers and Introduction to Solving Equations
2.1 Introduction to Integers 29
2.2 Adding Integers 33
2.3 Subtracting Integer 37
2.4 Multiplying and Dividing Integers 41
2.5 Order of Operations 45
2.6 Solving Equations: The Addition and Multiplication Properties 49

Module 3 Solving Equations and Problem Solving
3.1 Simplifying Algebraic Expressions 53
3.2 Revisiting the Properties of Equality 57
3.3 Solving Linear Equations in One Variable 61
3.4 Using Linear Equations to Solve Problems 65

Module 4 Fractions and Mixed Numbers
4.1 Introduction to Fractions and Mixed Numbers 69
4.2 Factors and Simplest Form 73
4.3 Multiplying and Dividing Fractions 77
4.4 Adding and Subtracting Fractions 81
4.5 Complex Fractions and Review of Order of Operations 85
4.6 Operations on Mixed Numbers 89
4.7 Solving Equations Containing Fractions 93

Module 5 Decimals
5.1 Introduction to Decimals 97
5.2 Adding and Subtracting Decimals 101
5.3 Multiplying Decimals; Circumference 105
5.4 Dividing Decimals 109
5.5 Fractions, Decimals, and Order of Operations 113
5.6 Solving Equations Containing Decimals 117

Module 6 Ratios and Proportions
6.1 Ratios, Rates, and Unit Prices 121
6.2 Proportions 125
6.3 Proportions and Problem Solving 129

Copyright © 2014 Pearson Education, Inc.

6.4 Congruent and Similar Triangles 133
6.5 Square Roots and the Pythagorean Theorem 137

Module 7 Percent
7.1 Percents, Decimals, and Fractions.. 141
7.2 Solving Percent Problems with Equations 145
7.3 Solving Percent Problems with Proportions... 149
7.4 Applications of Percent 153
7.5 Percent and Problem Solving: Sales Tax, Commission, and Discount 157
7.6 Percent and Problem Solving: Interest 161

Module 8 Geometry and Measurement
8.1 Lines and Angles.. 165
8.2 Perimeter, Circumference, and Area 169
8.3 Volume and Surface Area 173
8.4 Linear Measurement... 177
8.5 Weight and Mass.. 181
8.6 Capacity 185
8.7 Time and Temperature 189

Module 9 Statistics
9.1 Mean, Median, and Mode 193
9.2 Histograms 197
9.3 Counting 201
9.4 Probability 205

Module 10 Real Numbers and Algebraic Expressions
10.1 The Real Number System 209
10.2 Adding and Subtracting Real Numbers.. 213
10.3 Multiplying and Dividing Real Numbers 217
10.4 Exponents and Order of Operations 221
10.5 Variables and Properties of Real Numbers 225
10.6 Simplifying Algebraic Expressions.. 229

Module 11 Linear Equations and Inequalities in One Variable
11.1 The Addition and Multiplication Properties of Equality...................................... 233
11.2 Solving Linear Equations in One Variable 237
11.3 Introduction to Problem Solving.. 214
11.4 Formulas... 245
11.5 Geometry and Uniform Motion Problem Solving 249
11.6 Percent and Mixture Problem Solving 253
11.7 Linear Inequalities in One Variable 257
11.8 Compound Inequalities; Absolute Value Equations and Inequalities.................. 261

Module 12 Graphs of Linear Equations and Inequalities in Two Variables
12.1 The Rectangular Coordinate System.. 265

12.2 Graphing Linear Equations in Two Variables 269
12.3 Slope 273
12.4 Equations of Lines 277
12.5 Linear Inequalities in Two Variables 281

Module 13 Systems of Linear Equations and Inequalities
13.1 Solving Systems of Linear Equations by Graphing 285
13.2 Solving Systems of Linear Equations by Substitution 289
13.3 Solving Systems of Linear Equations by Elimination 293
13.4 Applications of Linear Systems 297
13.5 Systems of Linear Inequalities 301
13.6 Systems of Linear Equations in Three Variables 305

Module 14 Exponents and Polynomials
14.1 Exponents 309
14.2 Introduction to Polynomials 313
14.3 Adding and Subtracting Polynomials 317
14.4 Multiplying Polynomials 321
14.5 Special Products 325
14.6 Negative Exponents and Scientific Notation 329
14.7 Dividing Polynomials 333
14.8 Polynomials in Several Variables 337

Module 15 Factoring Polynomials
15.1 Greatest Common Factor and Factoring by Grouping 341
15.2 Factoring Trinomials of the Form $x^2 + bx + c$ 345
15.3 Factoring Trinomials of the Form $ax^2 + bx + c$ Using Trial and Error 349
15.4 Factoring Trinomials of the Form $ax^2 + bx + c$ Using the ac Method 353
15.5 Factoring Special Forms 357
15.6 A General Factoring Strategy 361
15.7 Solving Polynomial Equations by Factoring 365
15.8 Applications of Quadratic Equations 369

Module 16 Rational Expressions and Equations
16.1 Simplifying Rational Expressions 373
16.2 Multiplying and Dividing Rational Expressions 377
16.3 Least Common Denominators 381
16.4 Adding and Subtracting Rational Expressions 385
16.5 Complex Rational Expressions 389
16.6 Solving Rational Equations 393
16.7 Applications of Rational Equations 397
16.8 Variation 401

Module 17 Introduction to Functions
17.1 Relations and Functions 405
17.2 Function Notation and the Algebra of Functions 409

Copyright © 2014 Pearson Education, Inc.

17.3 Graphs of Functions and Their Applications 413

Module 18 Radicals and Rational Exponents
18.1 Radical Expressions 417
18.2 Radical Functions 421
18.3 Rational Exponents and Simplifying Radical Expressions 425
18.4 Operations with Radicals 429
18.5 Radical Equations and Models 433
18.6 Complex Numbers 437

Module 19 Quadratic Equations and Functions; Circles
19.1 Solving Quadratic Equations 441
19.2 Quadratic Functions and Their Graphs 443
19.3 Applications and Modeling of Quadratic Functions 449
19.4 Circles 453
19.5 Polynomial and Rational Inequalities 457

Module 20 Exponential and Logarithmic Functions and Equations
20.1 Transformations of Functions 461
20.2 Composite and Inverse Functions 465
20.3 Exponential Functions 469
20.4 The Natural Exponential Function 473
20.5 Logarithmic Functions 477
20.6 Properties of Logarithms 481
20.7 Exponential and Logarithmic Equations 485
20.8 Applications of Exponential and Logarithmic Functions 489

Module 21 Conic Sections
21.1 The Parabola 493
21.2 The Ellipse 497
21.3 The Hyperbola 501

Module 22 Sequences and Series
22.1 Introduction to Sequences and Series 505
22.2 Arithmetic Sequences and Series 509
22.3 Geometric Sequences and Series 513
22.4 The Binomial Theorem 517

Module 23 Additional Topics
23.1 Synthetic Division 521
23.2 Mean, Median, and Mode 523
23.3 Determinants and Cramer's Rule 527

Topic 1.1 Guided Notebook

Topic 1.1 Study Tips for This Course

Topic 1.1 Objective 1: Prepare for This Course

Write down the six tips that can help you have a successful start to your course. In your own words, provide a brief explanation for each tip.

1.

2.

3.

4.

5.

6.

Click on the **template** link found in tip number four on page 1.1-3. Print this document and fill out your schedule for the semester. Include work hours, classes, study time and other commitments. Study your schedule carefully and determine if you will have enough time for everything you have planned. Did you remember to leave yourself some leisure time?

Topic 1.1

Topic 1.1 Objective 2: Study for Optimal Success

There are ten tips for studying effectively for your mathematics course. Choose five that are most important to you and list them below. In your own words, provide a brief explanation for each tip.

1.

2.

3.

4.

5.

Tip number 8 states that "practice does not make perfect". Explain what this means in your own words.

Topic 1.1 Objective 3: Use the eText Effectively

There are four steps this eText provides for the benefit of students (although your instructor may not use all of them). List the four steps below:

1.

2.

3.

4.

Watch the video tour of the eText components that will be available to you. The link can be found on page 1.1-10. After viewing the tour, describe what each component is and how you might use it to help you achieve success in your math class.

Things to Know

You Try It

Animations

Videos

Interactive Videos

Topic 1.1

Popup definitions

Audio links

Sticky notes

Highlighter

Topic 1.1 Objective 4: <u>Get Ready for an Exam</u>

Many students suffer from "test anxiety", especially in mathematics courses. Read through the nine tips that are designed to help you prepare for an exam. Choose four that are most important to you and list them below. In your own words, provide a brief explanation for each tip.

1.

2.

3.

4.

Topic 1.2 Guided Notebook

Topic 1.2 Introduction to Whole Numbers

Topic 1.2 Objective 1: Identify the Place Value Of a Digit in a Whole Number

Write down the set of **whole numbers**.

What do the three dots at the end of the list mean? What are the three dots called?

What are **periods**?

Identify the first four periods from right to left:

1.

2.

3.

4.

Example 1:
Study the solutions for Example 1 parts a – c on page 1.2-3 and record the answers below.

Identify the place value of the digit 6 in each whole number.

a. 87,962   b. 506,721   c. 160,942,328

Topic 1.2 Objective 2: Write Whole Numbers in Standard Form and Word Form

Write down the method for **Writing Whole Numbers in Word Form.**

Topic 1.2

Example 2:
Study the solutions for Example 2 parts a and b on page 1.2-4, and record the answers below. Complete parts c and d on your own and check your answers by clicking on the link. If your answers are incorrect, watch the video to find your error.

Write each whole number in word form.

a. 894   b. 23,715   c. 6,047,829   d. 164,302,000

Read and summarize the CAUTION statement on 1.2-5.

Write down the method for **Writing Whole Numbers Standard Form.**

Example 3:
Study the solutions for Example 3 parts a – d on page 1.2-6. Record the answers below.

Write each whole number in standard form.

a. Six hundred twelve.

b. Eighteen thousand, two hundred fifty-seven.

c. One billion, three hundred seventy-two million, five hundred thirteen thousand, six hundred ninety-eight.

d. Two hundred sixty-nine million, five hundred thousand.

Read and summarize the second CAUTION statement on 1.2-7.

Topic 1.2

Topic 1.2 Objective 3: Change Whole Numbers from Standard Form to Expanded Form

Define the **expanded form** of a number.

Example 4:
Study the solution for Example 4 part a on page 1.2-8, and record the answer below. Complete part b on your own and check your answer by clicking on the link.

Write each whole number in expanded form.

a. 32,589

b. 4,520,708

Topic 1.2 Objective 4: Use Inequality Symbols to Compare Whole Numbers

Draw a number line showing the first 10 whole numbers. Graph the number 4 by placing a solid circle at its location on the number line.

What is the symbol for *greater than*? Use it correctly in a number sentence comparing two whole numbers.

What is the symbol for *less than*? Use it correctly in a number sentence comparing two whole numbers.

Example 5:
Study the solutions for Example 5 parts a and b on page 1.2-10. Record the answers below.

Fill in the blank with an inequality symbol, < or >, to make a true comparison statement.

a. 6 _____ 9

b. 15 _____ 12

Summarize the TIP found on page 1.2-10.

Topic 1.2

Topic 1.2 Objective 5: Round Whole Numbers

What does it mean to **round a whole number?**

Write down the steps for **Rounding a Whole Number to a Given Place Value.**

1.

2.

3.

Example 6:
Study the solutions for Example 6 parts a and b on page 1.2-12, and record the answers below. Complete parts c and d on your own and check your answers by clicking on the link. If your answers are incorrect, watch the video to find your error.

Round each whole number to the given place value.

a. 643 to the nearest ten

b. 3765 to the nearest hundred

c. 314,861 to the nearest ten-thousand

d. 29,534 to the nearest thousand

Topic 1.2 Objective 6: Read Tables and Bar Graphs Involving Whole Numbers

What are **tables** and **graphs** often used for?

Example 9:
Study the solution for Example 9 part a on page 1.2-18. Complete parts b and c on your own and check your answers by clicking on the link. If your answers are incorrect, watch the video to find your error.

b. What is the life expectancy of a male born in 2000?

c. In what year(s) of birth shown would a female be expected to live to the age of 79?

Topic 1.3 Guided Notebook

Topic 1.3 Adding and Subtracting Whole Numbers; Perimeter

Read the list of "THINGS TO KNOW" and review any concepts you are unfamiliar with.

Topic 1.3 Objective 1: Add Whole Numbers

What is a **sum**?

What are **addends**?

Example 2:
Complete Example 2 on page 1.3-5 on your own. Check your answer by clicking on the link. If your answer is incorrect, watch the video to find your error.

Add: 5203 + 1781

What does the term **carrying** mean in an addition problem?

Example 3:
Study the solution for Example 3 on page 1.3-6.

Example 4:
Complete Example 4 on page 1.3-8 on your own. Check your answer by clicking on the link. If your answer is incorrect, watch the video to find your error.

Add: 237,603 + 91,829

Topic 1.3 Objective 2: Use Special Properties of Addition

Write down the **Identity Property of Addition.**

Write down the **Commutative Property of Addition.**

Topic 1.3

Write down the **Associative Property of Addition.**

Example 5:
Study the solutions for Example 5 parts a and b on page 1.3-10. Record the answers below.

a. Use the commutative property of addition to complete the statement.

$4 + 9 =$ _____

b. Use the associative property of addition to complete the statement.

$5 + (3 + 6) =$ _____

Read and summarize the CAUTION statement on 1.3-11.

Topic 1.3 Objective 3: Add Several Whole Numbers

What two properties state that several numbers can be added in any order or grouping?

Example 7:
Complete Example 7 on page 1.3-14 on your own. Check your answer by clicking on the link. If your answer is incorrect, watch the video to find your error.

Add: $142 + 83 + 2061 + 17$

Topic 1.3 Objective 4: Subtract Whole Numbers

What is **subtraction**?

Use the following subtraction sentence to identify the **minuend, subtrahend** and **difference.**

$8 - 3 = 5$

Topic 1.3

Write down the **Identity Properties of Subtraction**.

Read and summarize the CAUTION statement on 1.3-18

Example 10:
Study the solution for Example 10 part a on page 1.3-19, and record the answer below. Complete part b on your own and check your answer by clicking on the link. If your answer is incorrect, watch the video to find your error.

Subtract. Check by adding.

a. 67 − 35   b. 428 − 205

What does the term **borrow** mean in a subtraction problem?

Example 11:
Study the solution for Example 11 part a on page 1.3-21, and record the answer below. Complete parts b and c on your own and check your answers by clicking on the link. If your answers are incorrect, watch the video to find your error.

Subtract. Check by adding.

a. 234 − 68   b. 500 − 269   c. 3270 − 1433

1.3 Objective 5: Estimate Sums and Differences of Whole Numbers

How do you **estimate** a sum or difference?

1.3 Objective 6: Find the Perimeter of a Polygon

What is the **perimeter** of a polygon? How do you find the perimeter?

Topic 1.3

Read and summarize the CAUTION statement on 1.3-28.

Example 15:
Complete Example 15 on page 1.3-29 on your own. Check your answer by clicking on the link. If your answer is incorrect, watch the video to find your error.

Find the perimeter of the polygon.

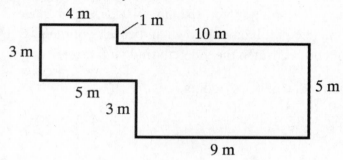

1.3 Objective 7: Solve Applications by Adding or Subtracting Whole Numbers

Study the key words in Figure 9 and Figure 10 on page 1.3-20.

Example 16:
Study the solutions for Example 16 parts a and b on page 1.3-31.

Example 18:
Complete Example 18 parts a and b on page 1.3-34 on your own. Check your answers by clicking on the link. If your answers are incorrect, watch the video to find your error.

John wants to fence off part of his land so his horses can graze. How much fencing will he need?

a. Round each value to the nearest hundred to estimate the total fencing needed.

b. Find the exact amount of fencing needed.

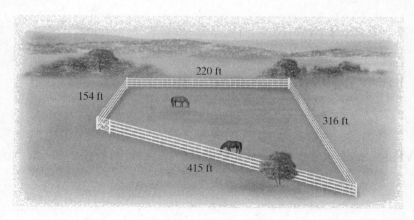

Topic 1.4 Guided Notebook

Topic 1.4 Multiplying Whole Numbers; Area

Read the list of "THINGS TO KNOW" and review any concepts you are unfamiliar with.

Topic 1.4 Objective 1: Use Special Properties of Multiplication

What is **multiplication**?

Use the following multiplication sentence to identify the **factors** and the **product**.
$$5 \times 7 = 35$$

Review basic multiplication facts by clicking on the *multiplication table* link on page 1.4-4.

Write down the **Commutative Property of Multiplication.**

Write down the **Associative Property of Multiplication.**

Write down the **Multiplication Property of Zero.**

Write down the **Identity Property of Multiplication.**

Topic 1.4 Objective 2: Use the Distributive Property

Write down the **Distributive Property.**

Read and summarize the CAUTION statement on 1.4-8.

Example 3:
Study the solutions for Example 3 parts a and b on page 1.4-9. Record the answers below.

Rewrite each statement using the distributive property.

a. $4(6+12)$ b. $(8-3)5$

Topic 1.4

Topic 1.4 Objective 3: Multiply Whole Numbers

Illustrate how to use the **distributive property** to multiply $4(17)$.

Example 4:
Study the solution for Example 4 part a on page 1.4-11 and record the answer below. Complete part b on your own and check your answer by clicking on the link. If your answer is incorrect, watch the video to find your error.

Multiply:

a. 243×8

b. 1472×6

Example 5:
Study the solutions for Example 5 parts a and b on page 1.4-13, and record the answers below. Complete part c on your own and check your answer by clicking on the link. If your answer is incorrect, watch the video to find your error.

Multiply:

a. 78
   × 34

b. 254
   × 429

c. 3402
   × 326

Topic 1.4 Objective 4: Estimate Products of Whole Numbers

Summarize the method for multiplying a number by multiples of 10, such as 10, 100 or 1000.

Example 7:
Complete Example 7 on page 1.4-21 on your own. Check your answer by clicking on the link. If your answer is incorrect, watch the video to find your error.

Estimate the product $26{,}159 \times 8733$ by first rounding the factors to the nearest thousand.

Topic 1.4 Objective 5: Find the Area of a Rectangle

What is the **area** of a figure?

Write down the formula for finding the area of a rectangle found on page 1.4-23.

Read and summarize the CAUTION statement on 1.4-23.

Example 8:
Study the solution for Example 8 part a on page 1.4-24, and record the answer below. Complete part b on your own and watch the video to confirm your answer.

Find the area of each rectangle.

a.

7 inches
18 inches

b.

80 meters
40 meters

Topic 1.4 Objective 6: Solve Applications by Multiplying Whole Numbers

Study the key words in Figure 11 on page 1.4-25.

Read and summarize the CAUTION statement on 1.4-25.

Topic 1.4

Example 10:
Complete Example 10 on page 1.4-26 on your own. Check your answer by clicking on the link. If your answer is incorrect, watch the video to find your error.

A 12-ounce can of Pepsi® contains 150 calories. How many total calories are there in a 24-pack of Pepsi? (*Source:* www.pepsicobeveragefacts.com)

Example 11:
Study the solution for Example 11 on page 1.4-27.

Example 12:
Study the solutions for Example 12 on page 1.4-27.

Example 13:
Complete Example 13 on page 1.4-28 on your own. Check your answer by clicking on the link. If your answer is incorrect, watch the video to find your error.

A youth group is planning to send 18 members to a rally in Washington, D.C. If each member attending the rally must pay $532 in airfare and $220 for food and lodging, what is the total cost of the trip for the entire group?

Topic 1.5

Topic 1.5 Guided Notebook

Topic 1.5 Dividing Whole Numbers

Read the list of "THINGS TO KNOW" and review any concepts you are unfamiliar with.

Topic 1.5 Objective 1: Divide Whole Numbers

What does it mean to **divide** a quantity?

Use the following division sentences to identify the **dividend**, **divisor**, and **quotient** in each.

$$24 \div 6 = 4 \qquad\qquad 6\overline{)24}^{\,4}$$

Example 1:
Study the solutions for Example 1 parts a and b on page 1.5-5, and record the answers below. Complete parts c and d on your own and check your answers by clicking on the link.

Find each quotient. Check by multiplying.

a. $45 \div 5$ b. $21/7$ c. $\dfrac{32}{8}$ d. $9\overline{)72}$

Write down the **Division Properties of 1.**

Example 3:
Study the solutions for Example 3 on page 1.5-7.

Write down the **Division Properties of 0.**

Read and summarize the CAUTION statement on page 1.5-8.

Topic 1.5

Example 4:
Study the solution for Example 4 on page 1.5-8.

When does a division problem have a **remainder**?

How do you check a division problem that has a remainder?

Example 5:
Study the solution for Example 5 part a on page 1.5-13, and record the answer below. Complete part b on your own and check your answer by clicking on the link. If your answer is incorrect, watch the video to find your error.

Divide. Check the result.

a. $6401 \div 7$ b. $16,542 \div 8$

Read and summarize the CAUTION statement on page 1.5-17.

Example 7:
Study the solutions for Example 7 on page 1.5-18.

Topic 1.5

Topic 1.5 Objective 2: Estimate Quotients of Whole Numbers

Summarize the method for dividing a number by multiples of 10, such as 10, 100 or 1000. Write down the **Division Shortcut.**

Example 8:
Study the solutions for Example 8 on page 1.5-21.

Read and summarize the first CAUTION statement on page 1.5-23.

When estimating a quotient, how is the division shortcut useful?

Read and summarize the second CAUTION statement on page 1.5-23.

Example 9:
Study the solution for Example 9 part a on page 1.5-24 and record the answer below. Complete part b on your own and check your answer by clicking on the link. If your answer is incorrect, watch the video to find your error.

Estimate each quotient.

a. $56,478 \div 82$ b. $715,809 \div 887$

Topic 1.5

Topic 1.5 Objective 3: Solve Applications by Dividing Whole Numbers

Study the key words in Figure 14 on page 1.5-25.

Example 10:
Study the solution for Example 10 on page 1.5-26, and record the answer below. View the popup box to see the work.

To throw a party, six friends will rent a hall for $450. If this rent is shared evenly among the friends, what is each friend's share of the rent?

Example 11:
Study the solution for Example 11 on page 1.5-27, and record the answer below. Watch the video for the detailed solution.

An executive secretary must charter busses to transport 965 workers from a conference hotel to a restaurant across town. If each bus carries 56 people, how many busses are needed? How many seats will be empty? (*Source:* www.newcharterbus.com)

Example 12:
Study the solution for Example 12 on page 1.5-28, and record the answer below.

During July 2011, the New York state debt was $283,477,740,095 when the population of New York was 19,466,875. Estimate the debt per citizen of New York at that time. (*Source:* US Debt Clock.org)

Topic 1.6

Topic 1.6 Guided Notebook

Topic 1.6 Exponents and Order of Operations

Read the list of "THINGS TO KNOW" and review any concepts you are unfamiliar with.

Topic 1.6 Objective 1: Use Exponential Notation

What is exponent notation used for?

Use the following product to write an **exponential expression**. Identify the base and exponent.

$$4 \cdot 4 \cdot 4 \cdot 4 \cdot 4$$

Example 1:
Study the solution for Example 1 part a on page 1.6-4, and record the answer below. Complete part b on your own and check your answer by clicking on the link. If your answer is incorrect, watch the video to find your error.

Write each of the following products using exponential notation.

a. $5 \cdot 5 \cdot 5$ b. $3 \cdot 3 \cdot 3 \cdot 3 \cdot 3 \cdot 3 \cdot 3$

Example 2:
Study the solution for Example 2 on page 1.6-5.

Topic 1.6 Objective 2: Evaluate Exponential Expressions

Explain what it means to **evaluate an expression**.

What is any number raised to the first power equal to?

Topic 1.6

Example 3:
Study the solutions for Example 3 on page 1.6-6, and record the answers below.

Evaluate each exponential expression.

a. 2^6 b. 10^4 c. 7^1

Read and summarize the CAUTION statement on 1.6-7.

Example 4:
Study the solution for Example 4 part a on page 1.6-7, and record the answer below. Complete part b on your own and check your answer by clicking on the link. If your answer is incorrect, watch the video to find your error.

Evaluate each exponential expression.

a. $4 \cdot 5^3$ b. $3^2 \cdot 10^3$

Topic 1.6 Objective 3: Simplify Expressions Using the Order of Operations

Write down the **Order of Operations.**

    1.

    2.

    3.

    4.

Topic 1.6

What is the tip for remembering the order of operations? Click on the popup on page 1.6-9.

Example 5:
Study the solutions for Example 5 parts a and b on page 1.6-9, and record the answers below. Complete part c on your own and check your answer by clicking on the link. If your answer is incorrect, watch the video to find your error.

Simplify each expression.

a. $7 + 5^2$ b. $70 \div 5 - 3 \cdot 2$ c. $10^2 \div 5 - 4 \cdot 2$

Example 6:
Complete Example 6 parts a and b on page 1.6-11 on your own. Check your answers by clicking on the link. If your answers are incorrect, watch the video to find your error.

Simplify each expression.

a. $22 + 5 - 3 + 6 - 4$ b. $48 \div 3 \cdot 2 - 5$

Example 7:
Study the solution for Example 7 part a on page 1.6-11, and record the answer below. Complete parts b and c on your own and check your answers by clicking on the link. If your answers are incorrect, watch the video to find your error.

Simplify each expression.

a. $(4^2 - 14) \cdot 3^2$ b. $3 \cdot (7 - 5)^2 + 4$ c. $28 \div 7 + \left[6 - 2^2\right]^3$

Example 8:
Study the solution for Example 8 on page 1.6-12.

Topic 1.6

Read and summarize the CAUTION statement on 1.6-14.

Example 9:
Study the solution for Example 9 on page 1.6-14.

Topic 1.6 Objective 4: <u>Find the Average of a List of Numbers</u>

Write down how you find the **average** of a list of numbers.

Example 10:
Study the solution for Example 10 on page 1.6-17 and record the answer below.

Find the average of 7, 15, 8, 23, 47, and 14.

Example 11:
Complete Example 11 on page 1.6-18 on your own. Check your answer by clicking on the link. If your answer is incorrect, watch the video to find your error.

Based on data from hotels.com, the four most expensive U.S. cities for hotel rooms during 2010 are given in the following table.

City	Daily Rate
New York	$194
Honolulu	$159
Boston	$155
Santa Barbara	$144

Find the average of these four values. (*Source:* www.hotels.com)

Topic 1.7

Topic 1.7 Guided Notebook

Topic 1.7 Introduction to Variables, Algebraic Expressions, and Equations

Read the list of "THINGS TO KNOW" and review any concepts you are unfamiliar with.

Topic 1.7 Objective 1: Evaluate Algebraic Expressions

What is a **variable**?

What is a **constant**?

Define **algebraic expression** and provide two examples.

Write down how to **Evaluate Algebraic Expressions.**

Example 1:
Study the solutions for Example 1 parts a and b on page 1.7-5, and record the answers below. Complete parts c and d on your own and check your answers by clicking on the link. If your answers are incorrect, watch the video to find your error.

Evaluate each algebraic expression for the given value of the variable.

a. $n+4$ for $n = 3$

b. $7a$ for $a = 9$

c. $x - 12$ for $x = 15$

c. $\dfrac{y}{4}$ for $y = 20$

Read and summarize the CAUTION statement on 1.7-7.

Topic 1.7

Example 3:
Study the solutions for Example 3 parts a and b on page 1.7-8, and record the answers below. Complete parts c and d on your own and check your answers by clicking on the link. If your answers are incorrect, watch the video to find your error.

Evaluate each expression for the given values of the variables.

a. $x^2 + y^3$ for $x = 5$ and $y = 2$

b. $5(m-n) - 14$ for $m = 21$ and $n = 13$

c. $\dfrac{w^2 - 5z}{z}$ for $w = 8$ and $z = 4$

d. $8x - 2xy$ for $x = 11$ and $y = 3$

Topic 1.7 Objective 2: Distinguish Between Expressions and Equations

Define **algebraic equation** and provide two examples.

Read and summarize the CAUTION statement on 1.7-10.

Example 4:
Study the solutions for Example 4 on page 1.7-11.

Topic 1.7 Objective 3: Determine if a Value is a Solution to an Equation

What does it mean to **solve an equation**?

What is a **solution**?

Example 5:
Study the solutions for Example 5 parts a and b on page 1.7-13, and record the answers below. Complete parts c and d on your own and check your answers by clicking on the link. If your answers are incorrect, watch the video to find your error.

Determine if the given value of the variable is a solution to the equation.

a. $6x - 5 = 43; x = 8$

b. $4a = a + 8; a = 3$

c. $3n - 7 = 2n + 5; n = 10$

d. $\dfrac{y+11}{2} = 2(y-1); y = 5$

Topic 1.7 Objective 4: Translate Word Phrases into Algebraic Expressions

Table 2 contains key words and phrases that indicate certain operations. Complete the table.

Addition	Subtraction	Multiplication	Division

Example 6:
Study the solutions for Example 6 on page 1.7-16.

Read and summarize the CAUTION statement on 1.7-18.

Topic 1.7

Example 7:
Study the solution for Example 7 part a on page 1.7-19, and record the answer below. Complete part b on your own and check your answer by clicking on the link. If your answer is incorrect, watch the video to find your error.

Write each word phrase as an algebraic expression. Use x to represent the unknown number.

a. The sum of 4 and triple a number　　　b. 8 times the difference of a number and 9

Topic 1.7 Objective 5: Translate Sentences into Equations

List 5 of the twelve key words or phrases that translate to an equal sign:

Example 8:
Study the solutions for Example 8 parts a and b on page 1.7-21, and record the answers below. Complete parts c and d on your own and check your answers by clicking on the link. If your answers are incorrect, watch the video to find your error.

Translate each sentence into an equation. Let x represent the unknown number.

a. Ten more than a number is 42.

b. A number divided by 5 will be 18.

c. Four times a number, decreased by 27, gives the number.

d. The sum of a number and 6 is the same as the difference of 12 and the number.

Topic 2.1

Topic 2.1 Guided Notebook

Topic 2.1 Introduction to Integers

Read the list of "THINGS TO KNOW" and review any concepts you are unfamiliar with.

Topic 2.1 Objective 1: Graph Integers on a Number Line

Write down the set of **integers**.

Read and summarize the CAUTION statement on 2.1-4.

Graph the integers -5, 2, and 0 by placing a solid circle at its location on the number line.

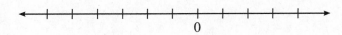

Topic 2.1 Objective 2: Use Inequality Symbols to Compare Integers

What is the symbol for *less than*? Use it correctly in a number sentence comparing two integers.

What is the symbol for *greater than*? Use it correctly in a number sentence comparing two integers.

Read and summarize the CAUTION statement on 2.1-6.

Example 2:
Study the solutions for Example 2 parts a – c on page 2.1-7. Record the answers below.

Fill in the blank with an inequality symbol, < or >, to make a true comparison statement.

a. −6 ____ 1   b. 0 ____ −8   c. −5 ____ −7

Topic 2.1

Topic 2.1 Objective 3: Find the Absolute Value of a Number

Define the **absolute value** of a number.

Read and summarize the CAUTION statement on 2.1-8.

Example 3:
Study the solutions for Example 3 parts a – c on page 2.1-8. Record the answers below.

Find each absolute value.

a. $|5|$ b. $|-4|$ c. $|0|$

Topic 2.1 Objective 4: Find the Opposite of a Number

Write down the definition of **opposites**.

Draw a number line illustrating 3 and its opposite.

Write down the method for **Finding the Opposite of a Number**.

Example 4:
Study the solutions for Example 4 parts a – c on page 2.1-11. Record the answers below.

Find the opposite of each number.

a. 16   b. −9   c. 0

Topic 2.1

One method of expressing $-a$ is "negative a". What is another method of expressing $-a$?

Write down the **Double-Negative Rule**.

Example 5:
Study the solutions for Example 5 parts a and b on page 2.1-12 and record the answers below. Complete parts c and d on your own and check your answers by clicking on the link. If your answers are incorrect, watch the video to find your error.

Simplify each expression.

a. $-(-20)$ b. $-|-8|$ c. $-|13|$ d. $|-(-6)|$

Topic 2.1 Objective 5: Solve Applications Involving Integers

Look at some applications involving integers.

Example 6:
Study the solutions for Example 6 parts a and b on page 2.1-14. Record the answers below.

In each situation, write the given number as an integer.

a. The TauTona gold mine in South Africa is currently the world's deepest mine at nearly 4 kilometers below the surface. (*Source:* National Geographic)

b. On September 13, 1922, the highest temperature ever recorded on Earth occurred in El Azizia, Libya at 136° above 0° Fahrenheit. (*Source:* National Climatic Data Center)

Topic 2.1

Example 7:
Study the solution for Example 7 on page 2.1-15, and record the answer below.

In Major League Baseball, integers are used to find a team's rank in its division by representing the number of games that the team is behind the front-runner. At the end of the 210 season, the Cincinnati Reds led the National League Central Division (NLC). The other teams in the NLC (and games behind) are the Chicago Cubs (-16), Houston Astros (-15), Milwaukee Brewers (-14), Pittsburgh Pirates (-34), and St. Louis Cardinals (-5). (*Source:* baseball-reference.com)

Use the information to rank the teams from best to worst in the NLC that season.

Rank	NLC Team	Games Behind the Front-runner
1	Cincinnati Reds	0
2		
3		
4		
5		
6		

Topic 2.2 Guided Notebook

Topic 2.2 Adding Integers

Read the list of "THINGS TO KNOW" and review any concepts you are unfamiliar with.

Topic 2.2 Objective 1: Add Two Integers with the Same Sign

Adding integers can be visualized on a number line. Describe how to add 2 and 3 using a number line.

Example 1:
Study the solutions for Example 1 parts a and b on page 2.2-4. Record the answers below.

Add.

a. $3+4$ 	b. $-1+(-5)$

Write down the steps for **Adding Two Numbers with the Same Sign.**

1.

2.

3.

Example 4:
Complete Example 4 parts a and b on page 2.2-7 on your own. Check your answers by clicking on the link. If your answers are incorrect, watch the video to find your error.

Add.

a. $16+8$ 	b. $-17+(-39)$

Topic 2.2

Topic 2.2 Objective 2: Add Two Integers with Different Signs

Example 5:
Study the solution for Example 5 part a on page 2.2-8, and record the answer below. Complete part b on your own and check your answer by clicking on the link. If your answer is incorrect, watch the video to find your error.

Add.

a. $2+(-5)$ b. $-3+(7)$

Write down the steps for **Adding Two Numbers with Different Signs.**

1.

2.

3.

Example 7:
Complete Example 7 parts a – c on page 2.2-11 on your own. Check your answers by clicking on the link. If your answers are incorrect, watch the video to find your error.

Add.

a. $-48+73$ b. $-35+35$ c. $28+(-94)$

What is another term for **opposites**?

Illustrate the sum of a number and its opposite.

Topic 2.2

Write down the procedure for **Adding Two Numbers.**

   1.

   2.

Example 8:
Study the solutions for Example 8 parts a and b on page 2.2-12, and record the answers below. Complete parts c and d on your own and check your answers by clicking on the link. If your answers are incorrect, watch the video to find your error.

Add.

a. $24 + (-37)$ b. $44 + 79$ c. $-27 + (-115)$ d. $-274 + (513)$

Topic 2.2 Objective 3: Add More Than Two Integers

To add more than two integers, perform the additions in order from _____ to _____.

What two properties allow us to add numbers in any order?

Example 10:
Study the solution to Example 10 on page 2.2-15. Record the answer below.

Add: $-16 + 49 + 26 + (-35)$

35

Topic 2.2

Topic 2.2 Objective 4: Solve Applications Involving Addition of Integers

Click on the popup box on page 2.2-16 and review the key words that mean addition. Write at least 3 down.

Example 12:
Study the solution to Example 12 on page 2.2-17. Record the answer below.

At 6 am on February 1, 2011, the temperature in Cheyenne, WY was $-18°F$. By noon, the temperature had increased by $23°F$. What was the temperature at noon? (*Source:* www.weatherunderground.com)

Example 13:
Complete Example 13 on page 2.2-17 on your own. Check your answer by clicking on the link. If your answer is incorrect, watch the video to find your error.

The following shows the point change in the Dow Jones Industrial Average (DJIA) for 5 days in 2011. What was the total point change for the 5 days? (*Source:* finance.yahoo.com)

July 19	July 20	July 21	July 22	July 25
201	−12	157	−44	−87

Topic 2.3 Guided Notebook

Topic 2.3 Subtracting Integers

Read the list of "THINGS TO KNOW" and review any concepts you are unfamiliar with.

Topic 2.3 Objective 1: Subtract Integers

If a and b represent two numbers, then $a - b =$ _____ (rewrite using equivalent addition).

Example 1:
Study the solution for Example 1 part a on page 2.3-4, and record the answer below. Complete part b on your own and check your answer by clicking on the link. If your answer is incorrect, watch the video to find your error.

Subtract.

a. $19 - 7$ b. $-6 - 9$

Review the **Double-Negative Rule** and write it down (view the popup box on page 2.3-5).

Example 2:
Study the solution for Example 2 part a on page 2.3-5, and record the answer below. Complete parts b and c on your own and check your answers by clicking on the link. If your answers are incorrect, watch the video to find your error.

Subtract.

a. $-7 - (-6)$ b. $10 - (-16)$ c. $25 - (-25)$

Topic 2.3

Topic 2.3 Objective 2: Add and Subtract Integers

If an expression contains both addition and subtraction, we change each _____ to _____ and find the _____.

Rewrite the given expression using only addition:

$7 + 23 - 18 + 4 - (-9) =$

Example 3:
Study the solution to Example 3 on page 2.3-6. Record the answer below.

Simplify: $5 - 8 + (-3) - (-6)$

Example 4:
Complete Example 4 on page 2.3-7 on your own. Check your answer by clicking on the link. If your answer is incorrect, watch the video to find your error.

Simplify: $-12 + 24 - (-15) - 8 + 10$

Topic 2.3 Objective 3: Solve Applications Involving Subtraction of Integers

Click on the popup box on page 2.3-8 and review the key words that mean subtraction. Write at least 3 down.

Topic 2.3

Read and summarize the CAUTION statement on 2.3-8.

Example 5:
Study the solutions for Example 5 part a on page 2.3-8, and record the answer below. Complete part b on your own and check your answer by clicking on the link. If your answer is incorrect, watch the video to find your error.

For each phrase, write an expression involving subtraction and simplify.

a. The difference of -4 and 20.

b. -18 subtracted from 35.

Example 6:
Study the solution to Example 6 on page 2.3-9. Record the answer below.

Brandi overdrew her checking account so her balance was -$24. The bank charged her a $45 overdraft fee and deducted that amount from her account. After the fee, what was her balance?

Topic 2.3

Example 7:

Complete Example 7 on page 2.3-10 on your own. Check your answer by clicking on the link. If your answer is incorrect, watch the video to find your error.

Isabella has $425 in her checking account. She withdraws $275 for her share of rent and withdraws $78 to pay her cell phone bill. She then deposits a payroll check in the amount of $317. How much money is now in her checking account?

Topic 2.4 Guided Notebook

Topic 2.4 Multiplying and Dividing Integers

Read the list of "THINGS TO KNOW" and review any concepts you are unfamiliar with.

Topic 2.4 Objective 1: Multiply Two Integers

Watch the animation on page 2.4-3 to develop the rules for multiplying two numbers.

Write down the rule for **Multiplying a Positive Number and a Negative Number.**

Write down the rule for **Multiplying Two Negative Numbers.**

Write down the steps for **Multiplying Two Numbers.**

Multiply the absolute values of the two factors to get the absolute value of the product. Determine the sign of the product using the following rules:

  1.

  2.

  3.

Topic 2.4

Example 3:
Complete Example 3 parts a – d on page 2.4-7 on your own. Check your answers by clicking on the link. If your answers are incorrect, watch the video to find your error.

Multiply:

a. $(-8)(11)$ b. $(-6)(0)$ c. $15 \cdot 7$ d. $-13 \cdot (-12)$

Topic 2.4 Objective 2: Multiply More Than Two Integers

When multiplying more than two integers, what might be helpful?

Example 5:
Study the solutions for Example 5 parts a and b on page 2.4-9, and record the answers below. Complete part c on your own and check your answer by clicking on the link. If your answer is incorrect, watch the video to find your error.

Multiply.

a. $-5(-2)(-7)$ b. $-3(-2)(-5)(-4)$ c. $-4(-3)(-2)(-6)(-1)$

Write down the summary for **Multiplying an Even or Odd Number of Negative Factors**.

Read and summarize the CAUTION statement on 2.4-10.

Topic 2.4

Topic 2.4 Objective 3: Evaluate Exponential Expressions Involving an Integer Base

Write down the summary for **Even and Odd Powers of a Negative Base.**

What is the difference between $(-4)^2$ and -4^2?

Read and summarize the CAUTION statement on 2.4-14.

Example 8:
Study the solutions for Example 8 parts a and b on page 2.4-14, and record the answers below. Complete parts c – e on your own and check your answers by clicking on the link. If your answers are incorrect, watch the video to find your error.

Evaluate each exponential expression.

a. $(-8)^2$ b. -2^4 c. -10^2

d. $(-4)^3$ e. -6^3

Topic 2.4 Objective 4: Divide Integers

Write down the steps for **Dividing Two Numbers.**

Divide the absolute values of the numbers to get the absolute value of the quotient. Determine the sign of the quotient using the following rules:

   1.

   2.

   3.

   4.

Topic 2.4

Read and summarize the CAUTION statement on 2.4-16.

Example 9:
Study the solutions for Example 9 parts a and b on page 2.4-17, and record the answers below. Complete parts c and d on your own and check your answers by clicking on the link. If your answers are incorrect, watch the video to find your error.

Divide.

a. $(-15) \div (-3)$
b. $\dfrac{-56}{7}$
c. $72/(-8)$
d. $\dfrac{0}{-12}$

Topic 2.4 Objective 5: <u>Solve Applications by Multiplying and Dividing Integers</u>

Click on the popup box on page 2.4-18 and review the key words that mean multiplication and the key words that mean division. Write at least 3 for each down.

Example 11:
Study the solution to Example 11 on page 2.4-19. Record the answer below.

Each month, Taima spends $75 more than she earns by making charges on her credit card. Use an integer to describe Taima's financial condition after 12 months.

Topic 2.5 Guided Notebook

Topic 2.5 Order of Operations

Read the list of "THINGS TO KNOW" and review any concepts you are unfamiliar with.

Topic 2.5 Objective 1: Use the Order of Operations with Integers

Review the **Order of Operations** and write the steps below.

1.

2.

3.

4.

Read and summarize the CAUTION statement on 2.5-3.

Example 2:
Complete Example 2 parts a – c on page 2.5-5 on your own. Check your answers by clicking on the link. If your answers are incorrect, watch the video to find your error.

Simplify.

a. $\dfrac{4-10}{-2}$ b. $13 - 2^2$ c. $-80 \div (-10) \cdot (-4)$

Topic 2.5

Example 4:
Complete Example 4 parts a – d on page 2.5-7 on your own. Check your answers by clicking on the link. If your answers are incorrect, watch the video to find your error.

Simplify.

a. $12-(-2)\cdot|-4|$

b. $(7-4)(3-8)$

c. $20-(6\cdot 4-1)$

d. $-4-12\div 2\times(-3)$

Example 6:
Complete Example 6 parts a and b on page 2.5-9 on your own. Check your answers by clicking on the link. If your answers are incorrect, watch the video to find your error.

Simplify.

a. $\left|(-3)^2-4\right|-2^3+6$

b. $-4\cdot 5^3-(4-2\cdot 3)$

Topic 2.5 Objective 2: Evaluate Algebraic Expressions Using Integers

Click on the popup box and review how to evaluate algebraic expressions. Write a brief summary below.

Topic 2.5

Example 7:
Study the solutions for Example 7 parts a and b on page 2.5-10, and record the answers below. Complete parts c and d on your own and check your answers by clicking on the link. If your answers are incorrect, watch the video to find your error.

Evaluate each algebraic expression for the given value of the variable.

a. $x+6$ for $x=-2$

b. $9-m$ for $m=17$

c. $-5z$ for $z=-4$

d. $\dfrac{p}{-2}$ for $p=18$

Example 9:
Study the solutions for Example 9 parts a and b on page 2.5-12, and record the answers below. Complete parts c and d on your own and check your answers by clicking on the link. If your answers are incorrect, watch the video to find your error.

Evaluate each algebraic expression for the given values of the variables.

a. $3x-y$ for $x=2$ and $y=-8$

b. $5(x+y)-z$ for $x=-5$, $y=7$ and $z=10$

c. $-2m^2+n^3$ for $m=-3$ and $n=4$

d. b^2-4ac for $a=7$, $b=2$, and $c=-1$

Topic 2.5 Objective 3: Translate Word Phrases Using Integers

Review the key words and phrases from Table 2 on page 2.5-14.

Topic 2.5

Example 10:
Study the solutions for Example 10 parts a and b on page 2.5-14, and record the answers below. Complete parts c and d on your own and check your answers by clicking on the link. If your answers are incorrect, watch the video to find your error.

Write each word phrase as an algebraic expression. Use x to represent the unknown number.

a. 9 more than the product of –4 and a number

b. The sum of –8 and a number, divided by the difference of the number and 11

c. 6 less than twice the sum of a number and –15

d. A number divided by –3, increased by twice the number

Read and summarize the CAUTION statement on 2.5-16.

Topic 2.6

Topic 2.6 Guided Notebook

Topic 2.6 Solving Equations: The Addition and Multiplication Properties

Read the list of "THINGS TO KNOW" and review any concepts you are unfamiliar with.

Topic 2.6 Objective 1: Determine if an Integer is a Solution to an Equation

Define the **solution** of an algebraic equation.

Example 1:
Study the solution for Example 1 part a on page 2.6-3, and record the answer below. Complete part b on your own and check your answer by clicking on the link. If your answer is incorrect, watch the video to find your error.

Determine if the given value of the variable is a solution to the equation.

a. $3x + 5 = -16$; -7

b. $-3n = 20 - n$; -5

Topic 2.6 Objective 2: Use the Addition Property of Equality to Solve Equations

Write down the definition for **equivalent equations**.

Write down the **Addition Property of Equality**.

Explain in words the *meaning* of this property.

Topic 2.6

Example 2:
Study the solution for Example 2 on page 2.6-7. Record the answer below.

Solve the equation $x - 5 = 7$.

Can subtraction be used in the Addition Property of Equality? Explain your answer.

Example 3:
Study the solution for Example 3 on page 2.6-8. Record the answer below.

Solve $24 = x + 10$.

Summarize the TIP found on page 2.6-9.

Example 4:
Study the solution for Example 4 part a on page 2.6-10, and record the answer below. Complete parts b and c on your own and check your answers by clicking on the link. If your answers are incorrect, watch the video to find your error.

Solve each equation.

a. $n - 3 = -7$ b. $2 = y + 9$ c. $-4 + m = -13$

Topic 2.6

Topic 2.6 Objective 3: Use the Multiplication Property of Equality to Solve Equations

Write down the **Multiplication Property of Equality**.

Explain in words the *meaning* of this property.

What is **A Helpful Fact When Multiplying and Dividing?**

Example 5:
Study the solution for Example 5 on page 2.6-14. Record the answer below.

Solve the equation $\frac{x}{2} = 3$.

Can division be used in the Multiplication Property of Equality? Explain your answer.

Topic 2.6

Example 6:
Study the solution for Example 6 on page 2.6-16. Record the answer below.

Solve $4x = 36$.

Example 7:
Study the solutions for Example 7 parts a and b on page 2.6-17, and record the answers below. Complete parts c and d on your own and check your answers by clicking on the link. If your answers are incorrect, watch the video to find your error.

Solve each equation.

a. $-3z = 21$ b. $-8 = \dfrac{w}{-5}$ c. $-96 = 16m$ d. $\dfrac{t}{-7} = 0$

Topic 3.1

Topic 3.1 Guided Notebook

Topic 3.1 Simplifying Algebraic Expressions

Read the list of "THINGS TO KNOW" and review any concepts you are unfamiliar with.

Topic 3.1 Objective 1: Identify Terms, Coefficients, and Like Terms

Write down the definition of a **term**.

What is a **variable term**? Give an example.

What is a **constant term**? Give an example.

Write down the definition of a **coefficient**.

Example 1:
Study the solutions for Example 1 parts a and b on page 3.1-5, and record the answers below. Complete part c on your own and check your answer by clicking on the link. If your answer is incorrect, watch the video to find your error.

List the terms of each expression. Then identify the coefficient of each term.

a. $12m - 7n$
b. $x^2 - 13x + 42$
c. $7x^3 - x^2 + x - 16$

Read and summarize the CAUTION statement on 3.1-6.

Write down the definition of **like terms**.

Topic 3.1 Objective 2: Combine Like Terms

What does it mean to **simplify the expression**?

Topic 3.1

Example 3:
Study the solutions for Example 3 parts a and b on page 3.1-8, and record the answers below. Complete part c on your own and check your answer by clicking on the link. If your answer is incorrect, watch the video to find your error.

Simplify each expression by combining like terms.

a. $12y - 10y$

b. $4xy + xy + 19xy$

c. $x^2 - 17x^2 + 6y^2 - 5y^2$

Which two properties are used when **collecting like terms**?

Example 5:
Complete Example 5 on page 3.1-11 on your own. Check your answers by clicking on the link. If your answers are incorrect, watch the video to find your error.

Simplify each algebraic expression by combining like terms.

a. $5x - 2x$

b. $6x^2 - 12x - 3x^2 + 4x$

c. $3z - 2z^2 + 7z^2$

d. $6x^2 + 2x + 4x + 3$

e. $-3x + 5 - y + x - 8$

Topic 3.1 Objective 3: Multiply Algebraic Expressions

Example 6:
Study the solutions for Example 6 parts a – c on page 3.1-12, and record the answers below.

Simplify each expression by multiplying.

a. $-9(5m)$

b. $4(-6y^2)$

c. $-13(-7a)$

Topic 3.1

Topic 3.1 Objective 4: Multiply Using the Distributive Property

Write down the distributive property.

Example 7:
Study the solutions for Example 7 parts a and b on page 3.1-15, and record the answers below. Complete part c on your own and check your answer by clicking on the link. If your answer is incorrect, watch the video to find your error.

Multiply using the distributive property.

a. $9(2x+5)$
b. $-8(10x+2y)$
c. $-10(3m-4)$

Topic 3.1 Objective 5: Simplify Algebraic Expressions

Write down the two steps that might be used to simplify an algebraic expression.

Example 8:
Study the solutions for Example 8 parts a and b on page 3.1-16, and record the answers below. Complete parts c and d on your own and check your answers by clicking on the link. If your answers are incorrect, watch the video to find your error.

Simplify each algebraic expression.

a. $7(x-2)+9$
b. $7(n+13)-(5n+21)$

c. $-2(2y-11)+9(y-1)$
d. $35-10(3t+8)$

Topic 3.2 Guided Notebook

Topic 3.2 Revisiting the Properties of Equality

Read the list of "THINGS TO KNOW" and review any concepts you are unfamiliar with.

Topic 3.2 Objective 1: Solve Equations Using the Addition Property of Equality

Write down the addition property of equality.

Example 1:
Study the solution for Example 1 on page 3.2-3, and record the answer below.

Solve the equation $x - 6 = -15$

Example 2:
Study the solution for Example 2 part a on page 3.2-4, and record the answer below. Complete part b on your own and check your answer by clicking on the link. If your answer is incorrect, watch the video to find your error.

Solve each equation.

a. $3x = 7 + 2x$

b. $-4n + 5 = -3n$

Topic 3.2

Example 3:
Complete Example 3 on page 3.2-6 on your own. Check your answer by clicking on the link. If your answer is incorrect, watch the video to find your error.

Solve $6y - 7 = 5y + 2$

Topic 3.2 Objective 2: Solve Equations Using the Multiplication Property of Equality

Write down the multiplication property of equality.

Example 4:
Study the solutions for Example 4 parts a and b on page 3.2-7, and record the answers below.

Solve each equation.

a. $\dfrac{x}{-3} = 12$

b. $4y = -32$

Topic 3.2 Objective 3: Solve Equations Using Both Properties of Equality

When both properties are applied, what is the addition property used for? What is the multiplication property used for?

Example 5:
Study the solutions for Example 5 parts a and b on page 3.2-9, and record the answers below. Complete parts c and d on your own and check your answers by clicking on the link. If your answers are incorrect, watch the video to find your error.

Solve each equation.

a. $9x - 4 = 68$

b. $\dfrac{t}{3} + 5 = -1$

c. $2 - 13n = -63$

d. $10 = -3x - 17$

Topic 3.2

Example 6:
Study the solution for Example 6 part a on page 3.2-11, and record the answer below. Complete part b on your own and check your answer by clicking on the link. If your answer is incorrect, watch the video to find your error.

Solve each equation.

a. $6x = -2x + 40$

b. $15n + 42 = 8n$

Example 7:
Complete Example 7 on page 3.2-12 on your own. Check your answers by clicking on the link. If your answers are incorrect, watch the video to find your error.

Solve each equation.

a. $5x - 8 = 3x + 12$

b. $16 - 5t = 19 - 2t$

Topic 3.3

Topic 3.3 Guided Notebook

Topic 3.3 Solving Linear Equations in One Variable

Read the list of "THINGS TO KNOW" and review any concepts you are unfamiliar with.

Topic 3.3 Objective 1: Identify Linear Equations in One Variable

Write the definition of a **linear equation in one variable**. Provide two examples.

Linear equations are also called _____ because the exponent on the variable is understood to be 1.

Example 1:
Study the solutions for Example 1 parts a – c on page 3.3-4, and record the answers below. Complete parts d – f on your own and check your answers by clicking on the link. If your answers are incorrect, watch the video to find your error.

Determine if each is a linear equation in one variable. If not, state why.

a. $-2x+5-7$

b. $3y+1=7$

c. $x^3 - 5x = 14$

d. $7a - 3b = 27$

e. $\dfrac{3}{x+2} = 7$

f. $9x + 20 = 7 - 4x$

Topic 3.3

Topic 3.3 Objective 2: Solve Linear Equations Involving Non-Simplified Expressions

If a linear equation is not simplified, what is done first?

Example 2:
Study the solution for Example 2 on page 3.3-5, and record the answer below. Include the **check.**

Solve: $7x - 8 + 3x = 5x - 13$

Read and summarize the CAUTION statement on 3.3-7.

Example 3:
Study the solution for Example 3 part a on page 3.3-7, and record the answer below. Complete part b on your own and check your answer by clicking on the link. If your answer is incorrect, watch the video to find your error.

Solve:

a. $6x + 5 - 4x = 8 - x + 9$ b. $3 + 6x + 15 = -7x - 2 + 3x$

Topic 3.3 Objective 3: Solve Linear Equations Involving Grouping Symbols

Example 4:
Study the solution for Example 4 part a on page 3.3-8, and record the answer below. Complete part b on your own and check your answer by clicking on the link. If your answer is incorrect, watch the video to find your error.

Solve:

a. $3(x+4) = 6$

b. $5(3-2x) + 4 = 2x - 5$

Record the steps in **A General Strategy for Solving Linear Equations in One Variable.**

    1.

    2.

    3.

    4.

    5.

Topic 3.3

Example 5:
Complete Example 5 on page 3.3-10 on your own. Check your answers by clicking on the link. If your answers are incorrect, watch the video to find your error.

Solve:

a. $5(x+7)=3(x-5)$

b. $3(2x-1)-7=10-2(x-2)$

Topic 3.4

Topic 3.4 Guided Notebook

Topic 3.4 Using Linear Equations to Solve Problems

Read the list of "THINGS TO KNOW" and review any concepts you are unfamiliar with.

Topic 3.4 Objective 1: Translate Word Statements into Equations

Review the key words for operations and the key words that translate to an equal sign.

Example 1:
Study the solutions for Example 1 parts a and b on page 3.4-3, and record the answers below. Complete parts c and d on your own and check your answers by clicking on the link. If your answers are incorrect, watch the video to find your error.

Write each word statement as an equation. Use x to represent the unknown number.

a. 14 more than a number is -20.

b. The difference of a number and 7 equals twice the number increased by 5.

c. The product of -4 and the sum of a number and 7 is the same as the number decreased by 10.

d. The quotient of a number and 5 results in -3 plus eight times the number.

Topic 3.4

Topic 3.4 Objective 2: Solve Problems Involving Numbers

What is the definition of a **mathematical model**?

Record the steps for the **Problem-Solving Strategy for Applications of Linear Equations.**

  1.

  2.

  3.

  4.

  5.

  6.

Example 2:
Study the solution for Example 2 on page 3.4-8. Record the answer below.

Three times a number, decreased by 4, equals the number increased by 2. Find the number.

Topic 3.4

Example 4:
Complete Example 4 on page 3.4-11 on your own. Check your answer by clicking on the link. If your answer is incorrect, watch the video to find your error.

Four times the difference of three times the number and 2, increased by 5 is the same as seven times the difference of the number and 4. Find the number.

Topic 3.4 Objective 3: Solve Problems Involving Perimeter and Area

Review the formulas for perimeter and area of a rectangle.

Example 5:
Study the solution for Example 5 on page 3.4-12. Record the answer below.

Example 8:
Complete Example 8 on page 3.4-17 on your own. Check your answer by clicking on the link. If your answer is incorrect, watch the video to find your error.

A photograph is 10 cm wide and 15 cm long. To fit in a picture frame, the picture length must be trimmed the same amount on both sides. If the final area is 110 square centimeters, how much should be trimmed from each side?

Topic 3.4 Objective 4: Solve Applications Using Linear Equations

Example 9:
Study the solution for Example 9 on page 3.4-19. Record the answer below.

During a blood drive, 12 more pints were collected the first day than on the second day. If the total number of pints collected over the two days was 58, how many pints were collected each day?

Example 10:
Complete Example 10 on page 3.4-22 on your own. Check your answer by clicking on the link. If your answer is incorrect, watch the video to find your error.

During Week 2 of the 2011 NFL season, Cam Newton had a total of 485 yards passing and running. If he passed for 8 yards more than 8 times his number of running yards, for how many yards did he pass? For how many yards did he run? (*Source*: espn.com)

Topic 4.1

Topic 4.1 Guided Notebook

Topic 4.1 Introduction to Fractions and Mixed Numbers

Read the list of "THINGS TO KNOW" and review any concepts you are unfamiliar with.

Topic 4.1 Objective 1: Identify the Numerator and Denominator of a Fraction

Write the definition of a **fraction**. Identify the parts of the fraction.

Topic 4.1 Objective 2: Represent Information with Fractions and Mixed Numbers

Example 2:
Study the solutions for Example 2 parts a and b on page 4.1-6, and record the answers below.

Write a fraction to represent the shaded portion of each figure.

a. 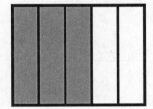   b.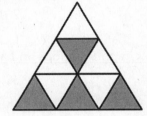

Example 5:
Complete Example 5 on page 4.1-9 on your own. Check your answer by clicking on the link. If your answer is incorrect, watch the video to find your error.

A prealgebra class has 15 male students and 13 female students. What fraction of the class is male?

Write the definition of a **proper fraction**. Provide one example.

Write the definition of an **improper fraction**. Provide one example.

Topic 4.1

Write the definition of a **mixed number**. Identify the parts of the mixed number.

Example 7:
Study the solution for Example 7 part a on page 4.1-13, and record the answer below. Complete part b on your own and check your answer by clicking on the link. If your answer is incorrect, watch the video to find your error.

For parts a and b, write both an improper fraction and a mixed number to represent the shaded portion in each case.

a. 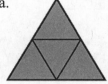   b.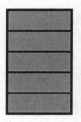

Topic 4.1 Objective 3: Graph Fractions and Mixed Numbers on a Number Line

Define **unit distance**.

When graphing fractions on a number line, the _____ tells us how many equal parts to divide the _____ into. The _____ tells us the position of the fraction within these parts.

Example 9:
Study the solution for Example 9 part a on page 4.1-16, and record the answer below. Complete parts b and c on your own and check your answers by clicking on the link. If your answers are incorrect, watch the video to find your error.

Graph each fraction on a number line.

a. $\frac{2}{3}$ b. $\frac{1}{5}$ c. $\frac{5}{6}$

Topic 4.1

The graphs of all proper fractions are between _____ and _____.

Improper fractions have values that are _____ than or _____ to 1.

Example 10:
Study the solution for Example 10 part a on page 4.1-18, and record the answer below. Complete parts b and c on your own and check your answers by clicking on the link. If your answers are incorrect, watch the video to find your error.

Graph each improper fraction on a number line.

a. $\dfrac{7}{4}$ 　　　　b. $\dfrac{5}{2}$ 　　　　c. $\dfrac{3}{3}$

Describe how to graph a mixed number.

Example 11:
Study the solution for Example 11 part a on page 4.1-20, and record the answer below. Complete part b on your own and check your answer by clicking on the link. If your answer is incorrect, watch the video to find your error.

Graph each mixed number on a number line.

a. $1\dfrac{3}{5}$ 　　　　b. $2\dfrac{1}{4}$

Topic 4.1 Objective 4: Write Improper Fractions as Mixed Numbers or Whole Numbers

Record the steps for **Writing an Improper Fraction as a Mixed Number or Whole Number.**

　　1.

　　2.

Topic 4.1

Example 12:
Study the solutions for Example 12 parts a and b on page 4.1-22, and record the answers below. Complete part c on your own and check your answer by clicking on the link. If your answer is incorrect, watch the video to find your error.

Write each improper fraction as a mixed number or whole number.

a. $\dfrac{74}{9}$ b. $\dfrac{126}{7}$ c. $\dfrac{185}{12}$

Topic 4.1 Objective 5: Write Mixed Numbers as Improper Fractions

Record the steps for **Writing a Mixed Number as an Improper Fraction.**

1.

2.

3.

Example 13:
Study the solution for Example 13 part a on page 4.1-24, and record the answer below. Complete part b on your own and check your answer by clicking on the link. If your answer is incorrect, watch the video to find your error.

Write each mixed number as an improper fraction.

a. $3\dfrac{2}{5}$ b. $12\dfrac{9}{16}$

Topic 4.2

Topic 4.2 Guided Notebook

Topic 4.2 Factors and Simplest Form

Read the list of "THINGS TO KNOW" and review any concepts you are unfamiliar with.

Topic 4.2 Objective 1: List the Factors of a Number

What does it mean to **factor** a nonzero whole number?

Write down the **Divisibility Tests for 2, 3, 5, and 10**.

Example 1:
Study the solution for Example 1 part a on page 4.2-4, and record the answer below. Complete parts b and c on your own and check your answers by clicking on the link. If your answers are incorrect, watch the video to find your error.

List the factors for each number.

a. 36   b. 64   c. 120

Topic 4.2 Objective 2: Find the Prime Factorization of a Number

Write the definition of a **prime number**.

Write the definition of a **composite number**.

Read and summarize the CAUTION statement on 4.2-8.

Topic 4.2

Record the steps for **Finding the Prime Factorization of a Composite Number.**

1.

2.

Read and summarize the CAUTION statement on 4.2-9.

Example 2:
Study the solution for Example 2 part a on page 4.2-10, and record the answer below. Complete parts b and c on your own and check your answers by clicking on the link. If your answers are incorrect, watch the video to find your error.

Find the prime factorization of each composite number.

a. 42　　　　　　　　　　b. 60　　　　　　　　　　c. 252

Topic 4.2 Objective 3: Find the Greatest Common Factor

List the factors of 30 and 42, then underline the largest common factor, which is called the
_____.

Example 4:
Study the solutions for Example 4 parts a and b on page 4.2-15, and record the answers below. Complete part c on your own and check your answer by clicking on the link. If your answer is incorrect, watch the video to find your error.

Find the GCF for each list of numbers.

a. 40, 60　　　　　　　　b. 108, 180　　　　　　　　c. 252, 420, 980

Topic 4.2

Record the steps for **Finding the Greatest Common Factor**.

    1.

    2.

    3.

Example 5:
Study the solution for Example 5 part a on page 4.2-17, and record the answer below. Complete parts b and c on your own and check your answers by clicking on the link. If your answers are incorrect, watch the video to find your error.

Find the GCF.

a. 225, 945   b. $4a^2b^4c, 6a^7b^3$ c. $12m^2n^5, 8m^3n^3, 20m^4n^3$

Topic 4.2 Objective 4: Determine If Two Fractions are Equivalent

Write the definition of **equivalent fractions**.

Example 6:
Study the solution for Example 6 part a on page 4.2-21, and record the answer below.

Determine if the pair of fractions is equivalent.

a. $\dfrac{3}{7}$ and $\dfrac{27}{63}$

Topic 4.2 Objective 5: Write Equivalent Fractions

Write down the **Property of Equivalent Fractions**.

Topic 4.2

Example 7:
Study the solution for Example 7 part a on page 4.2-23, and record the answer below. Complete part b on your own and check your answer by clicking on the link. If your answer is incorrect, watch the video to find your error.

Write the equivalent fraction with the given numerator or denominator.

a. $\dfrac{4}{5}$; denominator 35

b. $\dfrac{30}{54}$; numerator 5

Topic 4.2 Objective 6: Write Fractions in Simplest Form

Write the definition for the **simplest form for fractions**.

Summarize the TIP found on page 4.2-28.

Record the steps for **Writing a Fraction in Simplest Form**.

    1.

    2.

    3.

Example 10:
Study the solution for Example 10 part a on page 4.2-30, and record the answer below. Complete part b on your own and check your answer by clicking on the link. If your answer is incorrect, watch the video to find your error.

Write each fraction in simplest form.

a. $\dfrac{30x^4}{9x^2}$

b. $\dfrac{42x^2 y^3}{14x^3 y}$

Topic 4.3

Topic 4.3 Guided Notebook

Topic 4.3 Multiplying and Dividing Fractions

Read the list of "THINGS TO KNOW" and review any concepts you are unfamiliar with.

Topic 4.3 Objective 1: Multiply Fractions

Record the rule for **Multiplying Fractions**.

Example 2:
Study the solution for Example 2 parts a and b on page 4.3-8, and record the answers below. Complete parts c and d on your own and check your answers by clicking on the link. If your answers are incorrect, watch the video to find your error.

Multiply and simplify.

a. $\dfrac{2}{7} \cdot \dfrac{3}{4}$ b. $\dfrac{3}{50} \cdot \dfrac{10}{21}$ c. $24 \cdot \dfrac{5}{8}$ d. $\dfrac{7}{15} \cdot \dfrac{55}{14}$

Summarize the TIP found on page 4.3-10.

Example 4:
Study the solution for Example 4 part a on page 4.3-12, and record the answer below. Complete part b on your own and check your answer by clicking on the link. If your answer is incorrect, watch the video to find your error.

Multiply and simplify. Assume variables do not cause any denominator to equal 0.

a. $\dfrac{x}{y} \cdot \dfrac{y^2}{x^3}$ b. $\dfrac{6m}{4n} \cdot \dfrac{5}{2m}$

Topic 4.3

Topic 4.3 Objective 2: Evaluate Exponential Expressions Involving Fraction Bases

Example 5:
Study the solution for Example 5 part a on page 4.3-14, and record the answer below. Complete part b on your own and check your answer by clicking on the link. If your answer is incorrect, watch the video to find your error.

a. $\left(\dfrac{2}{3}\right)^4$
b. $\left(-\dfrac{5}{8}\right)^2$

Topic 4.3 Objective 3: Find Reciprocals

What number is the additive identity? _____ What number is the multiplicative identity? _____

Write the definition of **reciprocal or multiplicative inverse**. Give one example.

What is the reciprocal of $\dfrac{a}{b}$?

Read and summarize the CAUTION statement on 4.3-17.

Topic 4.3 Objective 4: Divide Fractions

Write the method for **Dividing Fractions**.

Read and summarize the CAUTION statement on 4.3-21.

Example 8:
Study the solution for Example 8 part a on page 4.3-21, and record the answer below. Complete part b on your own and check your answer by clicking on the link. If your answer is incorrect, watch the video to find your error.

Divide and simplify.

a. $9 \div \left(-\dfrac{3}{8}\right)$
b. $\left(-\dfrac{8}{21}\right) \div \left(-\dfrac{6}{14}\right)$

Topic 4.3 Objective 5: Simplify Expressions by Multiplying and Dividing Fractions

Example 10:
Study the solution for Example 10 part a on page 4.3-25, and record the answer below. Complete part b on your own and check your answer by clicking on the link. If your answer is incorrect, watch the video to find your error.

Multiply and divide to simplify each expression. Assume variables do not cause any denominators to equal 0.

a. $\dfrac{5}{9} \div \dfrac{7}{12} \cdot \dfrac{3}{10}$

b. $\left(-\dfrac{35p}{6q}\right)\left(-\dfrac{q^2}{55p}\right) \div \left(-\dfrac{14pq}{11}\right)$

Topic 4.3 Objective 6: Solve Applications by Multiplying or Dividing Fractions

What key word means to multiply?

Example 11:
Study the solution to Example 11 on page 4.3-27. Record the answer below.

Haley received a "five-eighths scholarship," which pays $\dfrac{5}{8}$ of the tuition and fees to her local community college. If tuition and fees total $2632, what dollar amount will Haley's scholarship cover? How much additional money will she need in order to pay the rest of the tuition and fees?

Example 12:
Complete Example 12 on page 4.3-28 on your own. Check your answer by clicking on the link. If your answer is incorrect, watch the video to find your error.

A *filibuster* is an action, such as long speechmaking, that is sometimes used in the U.S. Senate to prevent votes. At least $\dfrac{3}{5}$ of the votes in the Senate are needed to break (or stop) a filibuster. If all 100 senators vote, how many votes are needed to stop a filibuster?

Topic 4.3

Example 13:
Complete Example 13 on page 4.3-29 on your own. Check your answer by clicking on the link. If your answer is incorrect, watch the video to find your error.

To grill burgers for a holiday gathering, Otis purchased 12 pounds of ground beef to make into patties. If each burger patty is to weigh $\frac{3}{8}$ pound, how many burgers can Otis make?

Topic 4.3 Objective 7: Find the Area of a Triangle

Label the base, height and a vertex of the triangle.

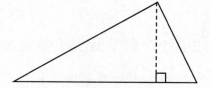

Watch the animation on page 4.3-31 that develops the formula for the area of a triangle.

Write down the formula for finding the **Area of a Triangle**.

Example 14:
Study the solution for Example 14 part a on page 4.3-32, and record the answer below. Complete part b on your own and check your answer by clicking on the link. If your answer is incorrect, watch the video to find your error.

Find the area of each triangle.

a.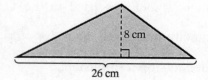

b.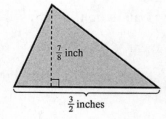

Topic 4.4

Topic 4.4 Guided Notebook

Topic 4.4 Adding and Subtracting Fractions

Read the list of "THINGS TO KNOW" and review any concepts you are unfamiliar with.

Topic 4.4 Objective 1: Add and Subtract Like Fractions

Like fractions are fractions with _____.

Unlike fractions are fractions with _____.

Complete the right side of the equation in **Adding Like Fractions**.

If a, b, and c are numbers, and $c \neq 0$, then $\dfrac{a}{c} + \dfrac{b}{c} =$

Example 1:
Study the solutions for Example 1 parts a and b on page 4.4-4, and record the answers below.

Add.

a. $\dfrac{3}{11} + \dfrac{5}{11}$

b. $\dfrac{4}{5} + \dfrac{3}{5}$

Example 2:
Study the solution for Example 2 part a on page 4.4-5, and record the answer below. Complete parts b and c on your own and check your answers by clicking on the link. If your answers are incorrect, watch the video to find your error.

Add.

a. $-\dfrac{3}{7} + \dfrac{4}{7}$

b. $-\dfrac{4}{9} + \left(-\dfrac{13}{9}\right)$

c. $-\dfrac{10}{3} + \dfrac{8}{3}$

Read and summarize the CAUTION statement on 4.4-6.

Topic 4.4

Complete the right side of the equation in **Subtracting Like Fractions**.

If a, b, and c are numbers, and $c \neq 0$, then $\dfrac{a}{c} - \dfrac{b}{c} =$

Example 4:
Study the solution for Example 4 part a on page 4.4-8, and record the answer below. Complete parts b – d on your own and check your answers by clicking on the link. If your answers are incorrect, watch the video to find your error.

Subtract. Simplify if necessary.

a. $\dfrac{7}{3} - \dfrac{5}{3}$
b. $-\dfrac{17}{6} - \dfrac{4}{6}$
c. $\dfrac{6}{7} - \dfrac{20}{7}$
d. $-\dfrac{3}{14} - \left(-\dfrac{5}{14}\right)$

Topic 4.4 Objective 2: Find the Least Common Denominator for Unlike Fractions

Write the definition for the **least common denominator (LCD)**.

Write the **Steps for Finding the LCD of a Group of Fractions by Prime Factorization**.
  1.

  2.

  3.

Example 6:
Study the solution for Example 6 part a on page 4.4-12, and record the answer below. Complete parts b and c on your own and check your answers by clicking on the link. If your answers are incorrect, watch the video to find your error.

Find the LCD of the given fractions by prime factorization.

a. $\dfrac{5}{18}$ and $\dfrac{7}{24}$
b. $\dfrac{11}{75}$ and $\dfrac{13}{120}$
c. $\dfrac{5}{6}, \dfrac{1}{21},$ and $\dfrac{23}{98}$

Topic 4.4

Topic 4.4 Objective 3: Add and Subtract Unlike Fractions

Write the **Steps for Adding or Subtracting Fractions**.

1.

2.

3.

4.

Example 7:
Study the solutions for Example 7 parts a – c on page 4.4-17, and record the answers below. Complete parts d – f on your own and check your answers by clicking on the link. If your answers are incorrect, watch the video to find your error.

Add or subtract. Simplify as necessary.

a. $\dfrac{1}{9} + \dfrac{5}{12}$

b. $\dfrac{11}{6} - \dfrac{1}{12}$

c. $\dfrac{3}{7} - 2$

d. $-\dfrac{11}{14} + \dfrac{5}{24}$

e. $\dfrac{7}{48} - \dfrac{8}{15}$

f. $\dfrac{10}{3} - \dfrac{7}{4} + \dfrac{1}{2}$

Topic 4.4 Objective 4: Add and Subtract Fractions with Variables

Example 9:
Study the solution for Example 9 part a on page 4.4-22, and record the answer below. Complete part b on your own and check your answer by clicking on the link. If your answer is incorrect, watch the video to find your error.

Add or subtract as indicated. Simplify if necessary.

a. $\dfrac{10}{3x} + \dfrac{2}{3x}$

b. $\dfrac{7}{6y^2} - \dfrac{2}{3y^2}$

Topic 4.4

Example 10:
Study the solutions for Example 10 parts a and b on page 4.4-24, and record the answers below.

Add or subtract as indicated. Simplify if necessary.

a. $\dfrac{7}{10x^2} + \dfrac{5}{2x}$

b. $\dfrac{1}{x} - \dfrac{2}{5}$

Example 11:
Complete Example 11 parts a – c on page 4.4-26 on your own. Check your answers by clicking on the link. If your answers are incorrect, watch the video to find your error.

Add or subtract as indicated. Simplify if necessary.

a. $\dfrac{5y}{3} - \dfrac{4}{3} + \dfrac{2y}{3}$

b. $\dfrac{a}{2} - \dfrac{6a}{7} + \dfrac{5}{4}$

c. $\dfrac{2}{3} - \dfrac{1}{6y} + \dfrac{3}{5}$

Topic 4.4 Objective 5: Solve Applications by Adding or Subtracting Fractions

Example 12:
Study the solution for Example 12 on page 4.4-27. Record the answer below.

A survey asked couples the month in which they became engaged. Of those surveyed, $\dfrac{8}{25}$ became engaged in either July or August. If $\dfrac{7}{50}$ became engaged in August, what fraction became engaged in July? (*Source*: bridalguide.com)

Topic 4.5 Guided Notebook

Topic 4.5 Complex Fractions and Review of Order of Operations

Read the list of "THINGS TO KNOW" and review any concepts you are unfamiliar with.

Topic 4.5 Objective 1: Simplify Complex Fractions

Write the definition of a **complex fraction**. Provide one example and identify the **minor fractions** and the **main fraction bar**.

Example 1:
Study the solution for Example 1 on page 4.5-4, and record the answer below.

Simplify the complex fraction.

$$\dfrac{\dfrac{3}{8}}{\dfrac{5}{4}}$$

Write down the steps of **Method 1 for Simplifying Complex Fractions**.

    1.

    2.

    3.

Example 2:
Study the solution for Example 2 on page 4.5-6, and record the answer below.

Use Method 1 to simplify the complex fraction.

$$\dfrac{\dfrac{5}{6} - \dfrac{3}{4}}{\dfrac{7}{9} + \dfrac{1}{4}}$$

85

Copyright © 2014 Pearson Education, Inc.

Topic 4.5

Write down the steps of **Method 2 for Simplifying Complex Fractions**.

1.

2.

3.

Example 4:

Complete Example 4 on page 4.5-10 on your own. Check your answer by clicking on the link. If your answer is incorrect, watch the video to find your error.

Use Method 2 to simplify the complex fraction.

$$\frac{\frac{5}{6}-\frac{3}{4}}{\frac{7}{9}+\frac{1}{4}}$$

Example 5:

Complete Example 5 parts a – c on page 4.5-12 on your own. Check your answers by clicking on the link. If your answers are incorrect, watch the video to find your error.

Use Method 1 or Method 2 to simplify each complex fraction. Assume variables do not cause any denominator to equal 0.

a. $\dfrac{\frac{7}{10}}{\frac{11}{15}-2}$
b. $\dfrac{\frac{5}{8}+\frac{17}{12}}{7}$
c. $\dfrac{\frac{x}{y}}{\frac{wx^2}{y^3}}$

Topic 4.5 Objective 2: Use the Order of Operations with Fractions

Review the **Order of Operations**.

1.

2.

3.

4.

Read and summarize the CAUTION statement on 4.5-13.

Example 6:
Study the solution for Example 6 part a on page 4.5-14, and record the answer below. Complete part b on your own and check your answer by clicking on the link. If your answer is incorrect, watch the video to find your error.

Simplify each expression.

a. $\dfrac{1}{2} + \dfrac{3}{4} \cdot \dfrac{1}{6}$

b. $\left(-\dfrac{5}{4}\right)^2 - \dfrac{7}{6}$

Topic 4.5

Example 7:
Study the solution for Example 7 part a on page 4.5-15, and record the answer below. Complete part b on your own and check your answer by clicking on the link. If your answer is incorrect, watch the video to find your error.

Simplify each expression.

a. $\dfrac{11}{10} - \dfrac{3}{10}\left(\dfrac{2}{3} - \dfrac{1}{6}\right)$

b. $\left(\dfrac{5}{6} - \dfrac{4}{9}\right)\left(\dfrac{1}{3} + \dfrac{7}{6}\right)$

Topic 4.5 Objective 3: Evaluate Algebraic Expressions Using Fractions

Click on the popup and review how to evaluate algebraic expressions.

Read and summarize the CAUTION statement on 4.5-18.

Example 9:
Complete Example 9 on page 4.5-18 on your own. Check your answer by clicking on the link. If your answer is incorrect, watch the video to find your error.

Evaluate $\dfrac{x+y}{z}$ for $x = -\dfrac{3}{5}$, $y = \dfrac{5}{6}$ and $z = \dfrac{7}{9}$.

Topic 4.6

Topic 4.6 Guided Notebook

Topic 4.6 Operations on Mixed Numbers

Read the list of "THINGS TO KNOW" and review any concepts you are unfamiliar with.

Topic 4.6 Objective 1: Multiply Mixed Numbers

Example 2:
Complete Example 2 parts a and b on page 4.6-5 on your own. Check your answers by clicking on the link. If your answers are incorrect, watch the video to find your error.

Multiply. Write the answer as a mixed number if possible.

a. $5 \cdot 3\frac{1}{4}$

b. $6\frac{2}{5} \cdot 4\frac{4}{9}$

Summarize the TIP found on page 4.6-6.

Topic 4.6 Objective 2: Divide Mixed Numbers

Example 3:
Study the solutions for Example 3 parts a and b on page 4.6-7, and record the answers below.

Divide. Write the answer as a mixed number if possible.

a. $2\frac{3}{4} \div \frac{8}{9}$

b. $\frac{3}{5} \div 5\frac{1}{7}$

Topic 4.6

Example 4:
Complete Example 4 parts a and b on page 4.6-9 on your own. Check your answers by clicking on the link. If your answers are incorrect, watch the video to find your error.

Divide. Write the answer as a mixed number if possible.

a. $4\dfrac{5}{9} \div 2\dfrac{1}{3}$
b. $7\dfrac{3}{5} \div 3$

Topic 4.6 Objective 3: Add Mixed Numbers

To add mixed numbers, first each mixed number could be converted into an improper fraction. What is another method?

Example 5:
Study the solution for Example 5 part a on page 4.6-10 and record the answer below. Complete part b on your own and check your answer by clicking on the link. If your answer is incorrect, watch the video to find your error.

Add. Write the answer as a mixed number if possible.

a. $6\dfrac{1}{3} + 2\dfrac{5}{12}$
b. $4\dfrac{1}{6} + 5\dfrac{3}{8}$

Explain how to simplify $4\dfrac{5}{3}$.

Topic 4.6

Example 7:
Complete Example 7 on page 4.6-15 on your own. Check your answer by clicking on the link. If your answer is incorrect, watch the video to find your error.

Add. Write your answer as a mixed number if possible.

$2\dfrac{3}{4} + 7 + 1\dfrac{4}{5}$

Topic 4.6 Objective 4: Subtract Mixed Numbers

When subtracting mixed numbers, when must you "borrow" from the whole number part?

Example 9:
Study the solution for Example 9 part a on page 4.6-19 and record the answer below. Complete part b on your own and check your answer by clicking on the link. If your answer is incorrect, watch the video to find your error.

Subtract. Write answers as a mixed number if possible.

a. $7\dfrac{3}{8} - \dfrac{5}{8}$

b. $4\dfrac{2}{5} - 1\dfrac{3}{4}$

Example 10:
Complete Example 10 on page 4.6-22 on your own. Check your answer by clicking on the link. If your answer is incorrect, watch the video to find your error.

Simplify. $8\dfrac{1}{3} + 1\dfrac{4}{9} - 3\dfrac{5}{18}$

Topic 4.6

Topic 4.6 Objective 5: Solve Applications Involving Mixed Numbers

Example 11:
Study the solution for Example 11 on page 4.6-23, and record the answer below.

The iPhone 4 is rectangular in shape and measures $4\frac{1}{2}$ inches long by $2\frac{3}{10}$ inches wide. Find its perimeter. (*Source*: www.apple.com)

Read and summarize the CAUTION statement on 4.6-24.

Topic 4.7 Guided Notebook

Topic 4.7 Solving Equations Containing Fractions

Read the list of "THINGS TO KNOW" and review any concepts you are unfamiliar with.

Topic 4.7 Objective 1: Determine If a Fraction Is a Solution to an Equation

What is a **solution** of an algebraic equation?

How is it determined if a given fraction is a solution to an equation?

Example 1:
Study the solution for Example 1 part a on page 4.7-4, and record the answer below. Complete part b on your own and check your answer by clicking on the link. If your answer is incorrect, watch the video to find your error.

Determine if the given fraction is a solution to the equation.

a. $3x - 1 = 9x - 5$; $\dfrac{2}{3}$

b. $-8y = 9 - 4y$; $-\dfrac{3}{4}$

Example 2:
Study the solutions for Example 2 parts a and b on page 4.7-5, and record the answers below.

Determine if the given value is a solution to the equation.

a. $\dfrac{2}{5}x - \dfrac{3}{10} = \dfrac{1}{2}$; 2

b. $\dfrac{5}{8}n + \dfrac{3}{4} = -\dfrac{5}{16}n$; $-\dfrac{4}{5}$

Topic 4.7

Topic 4.7 Objective 2: Use the Properties of Equality to Solve Linear Equations Involving Fractions

Click on the pop-ups to review the addition property of equality and the multiplication property of equality.

Example 3:
Study the solutions for Example 3 parts a and b on page 4.7-7, and record the answers below.

Solve each equation.

a. $x - \dfrac{3}{5} = \dfrac{1}{5}$

b. $\dfrac{9}{14} = m + \dfrac{3}{14}$

Example 6:
Study the solution for Example 6 part a on page 4.7-13, and record the answer below. Complete parts b – d on your own and check your answers by clicking on the link. If your answers are incorrect, watch the video to find your error.

Solve each equation.

a. $35 = -\dfrac{5}{7}k$

b. $\dfrac{w}{6} = \dfrac{5}{9}$

c. $20n = -\dfrac{4}{5}$

d. $60 = 8p$

Example 7:
Study the solution for Example 7 on page 4.7-15, and record the answer below.

Solve. $\dfrac{3}{10}x + \dfrac{7}{5} = \dfrac{1}{5}$

Topic 4.7

Topic 4.7 Objective 3: Solve Linear Equations by Clearing Fractions

Describe how to **clear the fractions** from an equation.

Example 8:
Study the solution for Example 8 part a on page 4.7-16, and record the answer below. Complete part b on your own and check your answer by clicking on the link. If your answer is incorrect, watch the video to find your error.

Solve each equation by first clearing the fractions.

a. $y + \dfrac{5}{4} = \dfrac{11}{12}$

b. $\dfrac{2}{15} = x - \dfrac{7}{10}$

Write down the steps for **A General Strategy for Solving Linear Equations in One Variable**.

1.

2.

3.

4.

5.

6.

Topic 4.7

Example 9:

Study the solution for Example 9 part a on page 4.7-18, and record the answer below. Complete part b on your own and check your answer by clicking on the link. If your answer is incorrect, watch the video to find your error.

Solve each equation by first clearing the fractions.

a. $\dfrac{x}{6} - \dfrac{x}{4} = -3$

b. $\dfrac{n}{5} = \dfrac{n}{15} + \dfrac{2}{3}$

Topic 5.1

Topic 5.1 Guided Notebook

Topic 5.1 Introduction to Decimals

Read the list of "THINGS TO KNOW" and review any concepts you are unfamiliar with.

Topic 5.1 Objective 1: Identify the Place Value of a Digit in a Decimal Number

Digits appearing in front of the decimal point form the _____ part of the decimal number. Digits appearing behind the decimal point form the _____ part (or _____ part) of the number.

Read and summarize the CAUTION statement on 5.1-4.

Example 1:
Study the solutions for Example 1 on page 5.1-4.

Topic 5.1 Objective 2: Write Decimals in Word Form

Record the steps for **Writing a Decimal Number in Words**.

1.

2.

3.

Example 2:
Study the solution for Example 2 parts a and b on page 5.1-6, and record the answers below. Complete parts c and d on your own and check your answers by clicking on the link. If your answers are incorrect, watch the video to find your error.

Write each decimal number in words.

a. 63.52   b. 259.3184   c. 0.436   d. 400.036

Copyright © 2014 Pearson Education, Inc.

Topic 5.1

Read and summarize the CAUTION statement on 5.1-7.

Topic 5.1 Objective 3: Change Decimals from Words to Standard Form

Record the steps for **Changing a Decimal from Words to Standard Form**.

    1.

    2.

    3.

Example 3:
Study the solution for Example 3 parts a and b on page 5.1-8, and record the answers below. Complete parts c and d on your own and check your answers by clicking on the link. If your answers are incorrect, watch the video to find your error.

Change each decimal to standard form.

a. Eight and fifteen hundredths

b. Four hundred sixty-three and five hundred seventy-one ten-thousandths

c. Two hundred eight thousandths

d. Two hundred and eight thousandths

Topic 5.1 Objective 4: Change Decimals into Fractions or Mixed Numbers

Record the steps for **Changing a Decimal into a Fraction or Mixed Number**.

    1.

    2.

    3.

    4.

Topic 5.1

Example 4:
Study the solution for Example 4 parts a and b on page 5.1-12, and record the answers below. Complete parts c and d on your own and check your answers by clicking on the link. If your answers are incorrect, watch the video to find your error.

Change each decimal into a fraction or mixed number. Simplify if necessary.

a. 0.69　　　　　b. 3.8　　　　　c. 0.375　　　　　d. 9.005

Topic 5.1 Objective 5: Graph Decimals on a Number Line

The place value of the last digit in a decimal determines the number of parts that the number line is divided into. To graph 1.4, how many parts is the interval between 1 and 2 divided into?

Example 5:
Study the solutions to Example 5 on page 5.1-17 and record the answers below.

Graph each decimal on a number line.

a. 0.6　　　　　b. −0.7　　　　　c. 1.76　　　　　d. −1.33

Topic 5.1 Objective 6: Compare Decimals

Where can a 0 be placed in a decimal so that it *will not* change the value of the number?

Read and summarize the CAUTION statement on 5.1-18.

Topic 5.1

Record the steps for **Comparing Two Decimals**.

1.

2.

3.

Example 6:
Study the solutions for Example 6 parts a – c on page 5.1-19, and record the answers below. Complete parts d – f on your own and check your answers by clicking on the link. If your answers are incorrect, watch the video to find your error.

Fill in the blank with < , > , or = to make a true comparison statement.

a. 0.531 ____ 0.529   b. 9.42 ____ 9.426   c. −0.65 ____ −0.69

d. −5.7 ____ −5.07   e. −0.7 ____ 0.3   f. 2.49 ____ 2.4900

Topic 5.1 Objective 7: Round Decimals to a Given Place Value

Record the steps for **Rounding a Decimal to the Right of the Decimal Point**.

1.

2.

Example 7:
Study the solution for Example 7 parts a and b on page 5.1-22, and record the answers below. Complete parts c and d on your own and check your answers by clicking on the link. If your answers are incorrect, watch the video to find your error.

Round each decimal to the given place value.

a. 23.769 to the nearest tenth

b. −435.8648 to the nearest hundredth

c. 18.369532 to the nearest thousandth

d. 361.208 to the nearest whole number

Topic 5.2

Topic 5.2 Guided Notebook

Topic 5.2 Adding and Subtracting Decimals

Read the list of "THINGS TO KNOW" and review any concepts you are unfamiliar with.

Topic 5.2 Objective 1: Add Positive Decimals

Write down the steps for **Adding Decimals**.

1.

2.

3.

Read and summarize the CAUTION statement on 5.2-5.

Example 2:
Study the solution for Example 2 part a on page 5.2-6, and record the answer below. Complete part b on your own and check your answer by clicking on the link. If your answer is incorrect, watch the video to find your error.

Add.

a. $57.62 + 28.741$

b. $3.019 + 0.085$

Example 3:
Study the solution for Example 3 on page 5.2-7. Record the answer below.

Add. $26 + 17.38$

Topic 5.2

Topic 5.2 Objective 2: Subtract Positive Decimals

Write down the steps for **Subtracting Decimals**.

1.

2.

3.

Summarize the TIP found on page 5.2-10.

Example 6:
Study the solution for Example 6 part a on page 5.2-11, and record the answer below. Complete parts b and c on your own and check your answers by clicking on the link. If your answers are incorrect, watch the video to find your error.

a. 94.02 − 35.64   b. 7.1371 − 0.95   c. 139 − 47.205

Topic 5.2 Objective 3: Estimate Sums and Differences of Decimals

For what reason is an estimate used?

Example 7:
Study the solution for Example 7 part a on page 5.2-13, and record the answer below. Complete part b on your own and check your answer by clicking on the link. If your answer is incorrect, watch the video to find your error.

Estimate the sum or difference by first rounding each number to the nearest ten.

a. 739.85 − 264.361   b. 28.6 + 154.608 + 77.39

Topic 5.2 Objective 4: Add and Subtract Negative Decimals

Example 8:
Study the solution for Example 8 part a on page 5.2-15, and record the answer below. Complete parts b – d on your own and check your answers by clicking on the link. If your answers are incorrect, watch the video to find your error.

Add or subtract as indicated.

a. $-37.48 + 74.63$

b. $-14.92 - 4.037$

c. $21 - 36.84$

d. $-46.72 - (-67.306)$

Topic 5.2 Objective 5: Add and Subtract Variable Expressions Involving Decimals

Click on the popup and review *combining like terms*.

Example 9:
Study the solution for Example 9 part a on page 5.2-18, and record the answer below. Complete part b on your own and check your answer by clicking on the link. If your answer is incorrect, watch the video to find your error.

a. $5.8 - 3.2x + 4.7x - 9.5$

b. $6.5y + 18.72 - 7.05y + 4.02y - 11$

Topic 5.2

Topic 5.2 Objective 6: Solve Applications by Adding or Subtracting Decimals

Example 10:
Study the solution for Example 10 on page 5.2-20. Record the answer below.

A speed test conducted by Engadget in February 2011 compared 3G download speeds of the Verizon iPhone 4 and the AT&T iPhone 4. The average download speed for the AT&T iPhone 4 was 2.702 Mbps (megabytes per second), while the average download speed for the Verizon iPhone 4 was 1.01 Mbps. What is the difference in download speed between the AT&T iPhone 4 and the Verizon iPhone 4? (*Source*: osxdaily.com)

Example 11:
Complete Example 11 on page 5.2-21 on your own. Check your answer by clicking on the link. If your answer is incorrect, watch the video to find your error.

TJ and some friends went out to dinner at a popular barbecue restaurant. The bill subtotal was $179.28 including tax. Because there were more than 6 people in their party, a tip was automatically added. If the added tip was $32.27, what was the final amount of the bill?

Topic 5.3 Guided Notebook

Topic 5.3 Multiplying Decimals; Circumference

Read the list of "THINGS TO KNOW" and review any concepts you are unfamiliar with.

Topic 5.3 Objective 1: Multiply Decimals

Write down the **Steps for Multiplying Decimals**.

1.

2.

Example 1:
Study the solution for Example 1 part a on page 5.3-4, and record the answer below. Complete parts b and c on your own and check your answers by clicking on the link. If your answers are incorrect, watch the video to find your error.

Multiply.

a. 4.2 x 7.6   b. 9.53 x 0.64   c. 12 x 1.006

Example 2:
Study the solution for Example 2 part a on page 5.3-5, and record the answer below. Complete part b on your own and check your answer by clicking on the link. If your answer is incorrect, watch the video to find your error.

Multiply.

a. $(5.78)(-3.2)$ b. $(-4.75)(-2.013)$

Topic 5.3

Topic 5.3 Objective 2: Multiply Decimals by Powers of 10

Study the pattern for multiplying by powers of 10 that are greater than or equal to 1.

Summarize the rule for **Multiplying Decimals by a Power of 10 Greater Than or Equal to 1.**

Study the pattern for multiplying by powers of 10 that are smaller than 1.

Summarize the rule for **Multiplying Decimals by a Power of 10 Less Than 1.**

Example 3:
Study the solution for Example 3 parts a and b on page 5.3-9, and record the answers below. Complete parts c and d on your own and check your answers by clicking on the link. If your answers are incorrect, watch the video to find your error.

Multiply.

a. 5.378 x 100

b. −18.735 x 0.1

c. 58.49 x 0.0001

d. $(-98.17)(-1000)$

Topic 5.3 Objective 3: Estimate Products of Decimals

Estimating the product of decimals is used to check the reasonableness of the answer. How else can estimating help?

Topic 5.3 Objective 4: Evaluate Exponential Expressions Involving Decimal Bases

Write the exponential expression $(3.5)^3$ in expanded form.

Topic 5.3

Example 5:
Study the solution for Example 5 part a on page 5.3-13, and record the answer below. Complete part b on your own and check your answer by clicking on the link. If your answer is incorrect, watch the video to find your error.

Evaluate each exponential expression.

a. $(4.3)^2$

b. $(-2.7)^3$

Topic 5.3 Objective 5: Find the Circumference of a Circle

What is the distance around a polygon called?

What is the distance around a circle called?

The value of π is estimated using a decimal or a fraction. What are two values that approximate π?

Write down the two formulas for the **Circumference of a Circle**.

Example 6:
Study the solution for Example 6 part a on page 5.3-17, and record the answer below. Complete part b on your own and check your answer by clicking on the link. If your answer is incorrect, watch the video to find your error.

Find the exact circumference of each circle in terms of π. Then approximate the circumference using 3.14 for π.

a.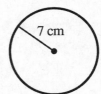

b.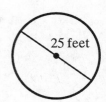

Topic 5.3

Topic 5.3 Objective 6: Solve Applications by Multiplying Decimals

Example 7:
Study the solution for Example 7 on page 5.3-19. Record the answer below.

According to the American Society of Plastic Surgeons, 2.4 million cosmetic procedures were performed on 30–39 year-olds in 2010. Write this number in expanded form. (*Source*: www.plasticsurgery.org)

Example 8:
Complete Example 8 on page 5.3-20 on your own. Check your answer by clicking on the link. If your answer is incorrect, watch the video to find your error.

Before returning a rental car, Leighton fills the gas tank. If the gasoline costs $3.179 per gallon and it takes 8.5 gallons to fill the tank, how much did it cost Leighton to fill the tank? Round the answer to two decimal places.

Topic 5.4

Topic 5.4 Guided Notebook

Topic 5.4 Dividing Decimals

Read the list of "THINGS TO KNOW" and review any concepts you are unfamiliar with.

Topic 5.4 Objective 1: Divide Decimals

Write down the steps for **Dividing a Decimal by a Whole Number**.

    1.

    2.

Summarize the TIP found on page 5.4-4.

Example 1:
Study the solution for Example 1 part a on page 5.4-4, and record the answer below. Complete part b on your own and check your answer by clicking on the link. If your answer is incorrect, watch the video to find your error.

Divide. Check the result.

a. $29.4 \div 6$ b. $-73.36 \div 14$

Example 2:
Study the solution for Example 2 part a on page 5.4-7, and record the answer below. Complete part b on your own and check your answer by clicking on the link. If your answer is incorrect, watch the video to find your error.

Divide. Check the result.

a. $22.8 \div 5$ b. $8.7 \div 12$

Topic 5.4

Write down the definition of a **terminating decimal**.

Write down the definition of a **repeating decimal**.

What is another way of expressing the decimal 4.57777…?

Read and summarize the CAUTION statement on 5.4-9.

Read and summarize the CAUTION statement on 5.4-11.

Example 4:
Study the solution for Example 4 part a on page 5.4-12, and record the answer below. Complete part b on your own and check your answer by clicking on the link. If your answer is incorrect, watch the video to find your error.

Divide. .

a. $-13 \div 8$

b. $-82 \div (-33)$

Record the steps for **Dividing Decimals**.

    1.

    2.

    3.

Topic 5.4

Example 5:
Study the solution for Example 5 parts a and b on page 5.4-15, and record the answers below. Complete parts c and d on your own and check your answers by clicking on the link. If your answers are incorrect, watch the video to find your error.

Divide.

a. $7.836 \div 0.8$

b. $-5.24 / 0.9$

c. $\dfrac{-0.105}{-0.875}$

d. $15.9 \div 0.33$

Summarize the TIP found on page 5.4-18.

Topic 5.4 Objective 2: Divide Decimals by Powers of 10

Study the pattern for dividing by powers of 10 that are greater than or equal to 1.

Summarize the rule for **Dividing Decimals by a Power of 10 Greater Than or Equal to 1.**

Study the pattern for dividing by powers of 10 that are smaller than 1.

Summarize the rule for **Dividing Decimals by a Power of 10 Less Than 1.**

Example 7:
Study the solutions for Example 7 parts a and b on page 5.4-22, and record the answers below. Complete parts c and d on your own and check your answers by clicking on the link. If your answers are incorrect, watch the video to find your error.

Divide.

a. $287.349 \div 100$

b. $92.594 / (-0.001)$

c. $\dfrac{-781.93}{-10,000}$

d. $\dfrac{3.91}{-0.00001}$

Topic 5.4

Topic 5.4 Objective 3: Estimate Quotients of Decimals

To obtain better estimates, what should be remembered?

Example 8:
Study the solution for Example 8 part a on page 5.4-24, and record the answer below. Complete part b on your own and check your answer by clicking on the link. If your answer is incorrect, watch the video to find your error.

Estimate each quotient by approximating the dividend and divisor with numbers that can be divided more easily.

a. $26.74 \div 2.8$

b. $394.632 \div 48.6$

Topic 5.4 Objective 4: Solve Applications by Dividing Decimals

Example 9:
Complete Example 9 on page 5.4-26 on your own. Check your answer by clicking on the link. If your answer is incorrect, watch the video to find your error.

To purchase a used car, Reagan borrowed $12,856.58. If the loan must be paid back in 48 equal monthly payments, what will be the monthly payment? Round to the nearest cent (hundredths place).

Topic 5.5

Topic 5.5 Guided Notebook

Topic 5.5 Fractions, Decimals, and Order of Operations

Read the list of "THINGS TO KNOW" and review any concepts you are unfamiliar with.

Topic 5.5 Objective 1: Change Fractions or Mixed Numbers into Decimals

Write down the method for **Changing a Fraction into a Decimal**.

Summarize the TIP found on page 5.5-3.

Example 1:
Study the solution for Example 1 on 5.5-3.

Example 2:
Study the solution for Example 2 on 5.5-5.

Example 3:
Study the solutions for Example 3 parts a and b on page 5.5-5, and record the answers below.

Change each fraction into an equivalent decimal.

a. $\dfrac{5}{6}$ b. $\dfrac{83}{99}$

Read and summarize the CAUTION statement on 5.5-7.

Example 4:
Complete Example 4 parts a – c on page 5.5-8 on your own. Check your answers by clicking on the link. If your answers are incorrect, watch the video to find your error.

Write each fraction as a decimal.

a. $\dfrac{9}{11}$ b. $\dfrac{7}{20}$ c. $\dfrac{19}{8}$

Topic 5.5

Summarize the TIP found on page 5.5-8.

Example 5:
Study the solution for Example 5 parts a and b on page 5.5-9. Record the answers below.

Change each mixed number into an equivalent decimal.

a. $3\frac{18}{25}$

b. $-5\frac{2}{3}$

Summarize the TIP found on page 5.5-10.

Example 6:
Study the solution for Example 6 part a on page 5.5-10, and record the answer below. Complete part b on your own and check your answer by clicking on the link. If your answer is incorrect, watch the video to find your error.

Find the decimal approximation for each fraction or mixed number, rounded to the given place value.

a. $\frac{3}{7}$; hundredths

b. $\frac{24}{19}$; thousandths

Topic 5.5 Objective 2: Compare Fractions and Decimals

Click on the popup and review the rules for comparing two decimals.

Example 7:
Study the solutions for Example 7 parts a – c on page 5.5-13, and record the answers below. Complete parts d – f on your own and check your answers by clicking on the link. If your answers are incorrect, watch the video to find your error.

Fill in the blank with <, >, or = to make a true comparison statement.

a. $\dfrac{3}{8}$ _____ 0.38

b. $\dfrac{5}{9}$ _____ 0.5

c. $\dfrac{11}{5}$ _____ 2.2

d. $\dfrac{7}{12}$ _____ $\dfrac{4}{7}$

e. $6\dfrac{7}{40}$ _____ 6.175

f. $0.\overline{56}$ _____ $\dfrac{17}{30}$

Topic 5.5 Objective 3: Use the Order of Operations with Decimals

Record the steps for simplifying numeric expressions using the **Order of Operations**.

1.

2.

3.

4.

Example 9:
Study the solution for Example 9 part a on page 5.5-18, and record the answer below. Complete part b on your own, and check your answer by clicking on the link. If your answer is incorrect, watch the video to find your error.

Simplify each using the order of operations.

a. $0.37(9.4-2.6)$

b. $(2.5)^2 + (1.4)^2$

Topic 5.5

Example 10:
Study the solution for Example 10 part a on page 5.5-19, and record the answer below. Complete part b on your own, and check your answer by clicking on the link. If your answer is incorrect, watch the video to find your error.

Simplify each using the order of operations.

a. $\dfrac{3}{5}(8.75) - 3.98$

b. $\dfrac{5 - (1.7)^2}{6(0.7) + 5.8}$

Topic 5.5 Objective 4: Evaluate Algebraic Expressions Using Decimals

Review how to evaluate algebraic expressions.

Topic 5.5 Objective 5: Find the Area of a Circle

Write down the definition for the **area** of a two-dimensional figure.

Write down the formula for the **Area of a Circle**.

Example 12:
Study the solution for Example 12 part a on page 5.5-22, and record the answer below. Complete part b on your own, and check your answer by clicking on the link. If your answer is incorrect, watch the video to find your error.

Find the exact area of each circle in terms of π. Then approximate the area using 3.14 for π.

a.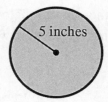

b.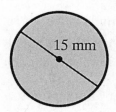

Topic 5.6

Topic 5.6 Guided Notebook
Topic 5.6 Solving Equations Containing Decimals

Read the list of "THINGS TO KNOW" and review any concepts you are unfamiliar with.

Topic 5.6 Objective 1: Determine If a Decimal Is a Solution to an Equation

Write the definition for a **solution** to an algebraic solution.

Example 1:
Study the solution for Example 1 part a on page 5.6-3, and record the answer below. Complete part b on your own and check your answer by clicking on the link. If your answer is incorrect, watch the video to find your error.

Determine if the given decimal is a solution to the equation.

a. $8x - 7 = 35.4$; 5.3

b. $1.5y + 4.8 = -3 + 2y$; -2.6

Topic 5.6 Objective 2: Use the Properties of Equality to Solve Linear Equations Involving Decimals

Click on the pop-up box and review the addition property of equality. Write both parts of the property below.

Click on the pop-up box and review the multiplication property of equality. Write both parts of the property below.

Topic 5.6

Example 2:
Study the solution for Example 2 part a on page 5.6-5, and record the answer below. Complete part b on your own and check your answer by clicking on the link. If your answer is incorrect, watch the video to find your error.

Solve each equation.

a. $x - 3.8 = 7$

b. $y + 4.3 = -2.3$

Example 3:
Study the solution for Example 3 part a on page 5.6-6, and record the answer below. Complete part b on your own and check your answer by clicking on the link. If your answer is incorrect, watch the video to find your error.

Solve each equation.

a. $\dfrac{x}{3} = -4.7$

b. $-2.4y = -16.8$

Example 5:
Complete Example 5 parts a and b on page 5.6-8 on your own. Check your answers by clicking on the link. If your answers are incorrect, watch the video to find your error.

Solve each equation.

a. $14.5x + 3 = 5x - 12.2$

b. $7z - 4.04 = 1.3z + 14.2$

Topic 5.6

Example 6:
Study the solution for Example 6 part a on page 5.6-9. Complete parts b and c on your own and check your answers by clicking on the link. If your answers are incorrect, watch the video to find your error.

Solve each equation.

b. $1.8(b+10) = 0.92 - b$

c. $2(m-3.7) = 4(m+5.2) - 44$

Topic 5.6 Objective 3: Solve Linear Equations by Clearing Decimals

Click on the pop-up box and review how to **clear the fractions** in a linear equation.

Example 7:
Study the solution for Example 7 part a on page 5.6-11, and record the answer below. Complete part b on your own and check your answer by clicking on the link. If your answer is incorrect, watch the video to find your error.

Solve each equation by first clearing the decimals.

a. $4x - 3.9 = -14.3$

b. $3.7y + 4.25 = 9.8$

Topic 5.6

Record the steps for **A General Strategy for Solving Linear Equations in One Variable**.

1.

2.

3.

4.

5.

6.

Topic 5.6 Objective 4: Use Linear Equations to Solve Applications Involving Decimals

Example 9:
Complete Example 9 on page 5.6-15 on your own. Check your answer by clicking on the link. If your answer is incorrect, watch the video to find your error.

Quan uses iTunes® to purchase some apps for $0.99 each and some songs for $1.29 each. If he purchased 15 more songs than apps and spent a total of $37.59, how many of each did he buy?

Topic 6.1

Topic 6.1 Guided Notebook

Topic 6.1 Ratios, Rates, and Unit Prices

Read the list of "THINGS TO KNOW" and review any concepts you are unfamiliar with.

Topic 6.1 Objective 1: Write Two Quantities as a Ratio

Define **ratio** and illustrate the three common notations for expressing them.

Read and summarize the CAUTION statement on 6.1-3.

Summarize the TIP found on page 6.1-5.

Record the method for **Simplifying a Ratio**.

Example 3:
Study the solution for Example 3 part a on page 6.1-7, and record the answer below. Complete part b on your own and check your answer by clicking on the link. If your answer is incorrect, watch the video to find your error.

a. Amazon's Kindle Fire tablet has a length of 190 mm and a width of 120 mm. Write the ratio of length to width as a fraction in simplest form. (*Source*: amazon.com)

b. A dietician recommends to a patient that he consume 125 grams of protein, 140 grams of fat, and 50 grams of carbohydrates each day. Write the ratio of carbohydrates to protein in simplest form.

Topic 6.1

Example 4:
Study the solution for Example 4 part a on page 6.1-9. Complete parts b and c on your own and check your answers by clicking on the link. If your answers are incorrect, watch the video to find your error.

Write each ratio in simplest form.

a. 4.2 to 2.8

b. $\frac{3}{4}$ to $\frac{1}{6}$

c. $3\frac{1}{4}$ to $2\frac{1}{2}$

Topic 6.1 Objective 2: Write Two Quantities as a Rate

Define **rate** and record the words other than 'to' that can be used to separate the quantities being compared.

Example 7:
Complete Example 7 parts a and b on page 6.1-16 on your own. Check your answers by clicking on the link. If your answers are incorrect, watch the video to find your error.

a. Tanner typed a 2000-word paper in 45 minutes. Write his typing rate as a simplified fraction.

b. During the 2011 season, Adrian Gonzalez of the Boston Red Sox hit 27 home runs in 630 at-bats. Write his home run to at-bats rate as a simplified fraction. (*Source*: espn.com)

Topic 6.1

Topic 6.1 Objective 3: Find a Unit Rate

Define **unit rate**.

Record the method for **Converting Rates to Unit Rates**.

Example 8:
Study the solutions for Example 8 on page 6.1-17.

Example 9:
Complete Example 9 parts a and b on page 6.1-18 on your own. Check your answers by clicking on the link. If your answers are incorrect, watch the video to find your error.

a. Using her DSL modem, Tameka can download a 360 MB file in 16 seconds. Find her modem's unit rate in MB per second.

b. The U.S. government often reports statistics as "per capita" (per person). If the population of the U.S. was 305,000,000 in 2009 and 1,982,500,000 pounds of peanuts were consumed during that year, find the unit rate of peanut consumption per person for 2009. In other words, find the amount of peanuts eaten per capita in the U.S. for 2009. (*Source*: U.S. Department of Agriculture)

Topic 6.1 Objective 4: Compare Unit Prices

Define a **unit price**.

Topic 6.1

Record the method for finding a **unit price**.

Example 10:
Study the solutions for Example 10 parts a and b on page 6.1-19, and record the answers below.

a. Piper used 60 MB of data on her smartphone beyond what is covered in her monthly plan. The cost for this additional data was $15. Find the unit rate in dollars per MB.

b. If a 12-ounce can of soda costs 75¢, what is the unit price in ¢ per ounce?

Example 12:
Complete Example 12 on page 6.1-21 on your own. Check your answer by clicking on the link. If your answer is incorrect, watch the video to find your error.

Grocery stores will often display unit prices on price labels so that shoppers can make comparisons quickly. The following price labels show that a 40-oz jar of a certain brand of creamy peanut butter sells for $6.38 and a 56-oz jar of the same type of peanut butter sells for $9.98 (the unit prices have been hidden). Find the unit prices and compare them to determine which is the better deal. Round to two decimal places if necessary.

Topic 6.2

Topic 6.2 Guided Notebook
Topic 6.2 Proportions

Read the list of "THINGS TO KNOW" and review any concepts you are unfamiliar with.

Topic 6.2 Objective 1: Write Proportions

Write the definition of a **proportion**. Provide an example.

Example 1:
Study the solutions for Example 1 parts a – c on page 6.2-4, and record the answers below.

Write each sentence as a proportion.

a. 2 is to 12 as 8 is to 48.

b. 424 miles is to 16 gallons as 318 miles is to 12 gallons.

c. 6 cups is to 4 cups as 1.5 cups is to 1 cup.

Summarize the TIP found on page 6.2-5.

Topic 6.2 Objective 2: Determine Whether Proportions Are True or False

One way to determine whether a proportion is true or false is to write both ratios in _____ and compare results.

Topic 6.2

Example 2:

Study the solutions for example 2 parts a and b on page 6.2-6, and record the answers below.

Write each ratio in simplest form to determine whether each proportion is true or false.

a. $\dfrac{15}{20} = \dfrac{18}{24}$

b. $\dfrac{45}{18} = \dfrac{24}{9}$

Write down a second method for determining whether a proportion is true or false.

Record the method for **Using Cross Products to Determine Whether Proportions are True or False.**

Example 3:

Study the solution for Example 3 part a on page 6.2-8, and record the answer below. Complete part b on your own and check your answer by clicking on the link. If your answer is incorrect, watch the video to find your error.

Use cross products to determine whether each proportion is true or false.

a. $\dfrac{8}{50} = \dfrac{12}{75}$

b. $\dfrac{4}{18} = \dfrac{9}{27}$

Topic 6.2

Example 4:
Study the solution for Example 4 part a on page 6.2-9, and record the answer below. Complete part b on your own and check your answer by clicking on the link. If your answer is incorrect, watch the video to find your error.

Use cross products to determine whether each proportion is true or false.

a. $\dfrac{3.5}{8} = \dfrac{5.4}{12}$

b. $\dfrac{\frac{3}{4}}{2\frac{1}{2}} = \dfrac{\frac{1}{2}}{1\frac{2}{3}}$

Topic 6.2 Objective 3: Solve Proportions

What does it mean to **solve the proportion**?

Write the steps for **Solving a Proportion**.

    1.

    2.

    3.

Topic 6.2

Example 5:
Study the solution for Example 5 part a on page 6.2-13, and record the answer below. Complete part b on your own and check your answer by clicking on the link. If your answer is incorrect, watch the video to find your error.

Solve each proportion.

a. $\dfrac{75}{x} = \dfrac{3}{8}$

b. $\dfrac{21}{14} = \dfrac{18}{y}$

Summarize the TIP found on page 6.2-13.

Example 8:
Study the solution for Example 8 part a on page 6.2-16, and record the answer below. Complete part b on your own and check your answer by clicking on the link. If your answer is incorrect, watch the video to find your error.

Solve each proportion. Write the solution as a simplified proper fraction or mixed number when applicable.

a. $\dfrac{\frac{1}{2}}{\frac{3}{4}} = \dfrac{x}{\frac{1}{6}}$

b. $\dfrac{m}{4\frac{1}{6}} = \dfrac{1\frac{4}{5}}{2\frac{3}{16}}$

Topic 6.3 Guided Notebook

Topic 6.3 Proportions and Problem Solving

Read the list of "THINGS TO KNOW" and review any concepts you are unfamiliar with.

Topic 6.3 Objective 1: Use Proportions to Solve Applications

Proportions can be used to solve a wide variety of application problems. Name three fields of study where proportions might be used.

Record the steps for the **Problem-Solving Strategy for Applications of Proportions**.

1.

2.

3.

4.

5.

6.

Topic 6.3

Example 1:
Study the solution for Example 1 on page 6.3-4, and record the answer below.

In 2011, researchers reported that the average 18-24 year-old sends and receives 110 text messages per day. At this rate, how many texts does the average 18-24 year-old send and receive in a 30-day month? (*Source*: Pew Research Center, April – May 2011)

Example 2:
Complete Example 2 on page 6.3-6 on your own. Check your answer by clicking on the link. If your answer is incorrect, watch the video to find your error.

To estimate the population of largemouth bass in a small lake, a conservation agent catches 120 bass, tags them, and releases them back into the lake. Some time later, the agent returns, catches a sample of 90 bass, and finds that 8 of them are tagged. Estimate the number of largemouth bass in the entire population.

Topic 6.3

Example 3:

Complete Example 3 on page 6.3-8 on your own. Check your answer by clicking on the link. If your answer is incorrect, watch the video to find your error.

On the given map showing part of the state of Illinois, an actual distance of 50 miles is represented by 2.5 centimeters. If the distance between Springfield and Chicago is 10.6 centimeters on the map, find the actual distance between the two cities.

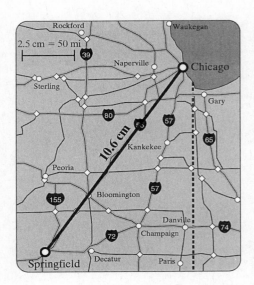

Topic 6.3

Example 4:

Complete Example 4 on page 6.3-10 on your own. Check your answer by clicking on the link. If your answer is incorrect, watch the video to find your error.

In his new hybrid car, Malik can travel 194 miles on 4.5 gallons of gas. How far can Malik travel on 12 gallons of gas? Round to the nearest mile.

Topic 6.4 Guided Notebook

Topic 6.4 Congruent and Similar Triangles

Read the list of "THINGS TO KNOW" and review any concepts you are unfamiliar with.

Topic 6.4 Objective 1: Identify the Corresponding Parts of Congruent Triangles

What is each corner point of a triangle called?

Name the triangle in Figure 1 three different ways.

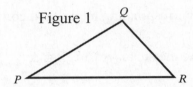

Figure 1

Identify the three angles and the three sides of the triangle in Figure 1.

Define **congruent**. What is the **congruence symbol**?

Define **corresponding angles** and **corresponding sides**.

Summarize the TIP found on page 6.4-5.

In the statement $m\angle A = m\angle D$, what does the m mean?

Topic 6.4

Example 1:

Study the solutions for Example 1 parts a – d on page 6.4-6, and record the answers below.

The two triangles shown below are congruent.

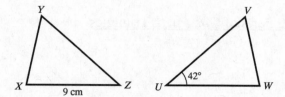

a. List the corresponding angles and corresponding sides.

b. Complete the sentence $\triangle XYZ \cong \triangle$ _____ .

c. Find the measure of angle Z.

d. Find the length of $\overline{UW}$.

Read and summarize the CAUTION statement on page 6.4-8.

Topic 6.4 Objective 2: Determine Whether Two Triangles Are Congruent

What is the **Side-Side-Side (SSS) Property of Congruence**? Draw and label a diagram.

What is the **Side-Angle-Side (SAS) Property of Congruence**? Draw and label a diagram.

Topic 6.4

What is the **Angle-Side-Angle (ASA) Property of Congruence**? Draw and label a diagram.

Example 2:
Study the solutions for Example 2 parts a and b on page 6.4-12.

Example 3:
Study the solution for Example 3 part a on page 6.4-13, and record the answer below. Complete part b on your own and check your answer by clicking on the link. If your answer is incorrect, watch the video to find your error.

Determine whether each pair of triangles is congruent. If congruent, state the applicable property of congruence (SSS, SAS, or ASA).

a.
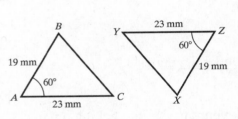

b.
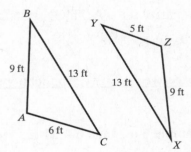

Topic 6.4 Objective 3: Identify the Corresponding Parts of Similar Triangles

Define **similar**. What is the **similarity symbol**?

Define **corresponding angles** and **corresponding sides** as they relate to similar triangles.

Example 4:
Study the solutions for Example 4 parts a and b on page 6.4-16.

135

Copyright © 2014 Pearson Education, Inc.

Topic 6.4

Topic 6.4 Objective 4: Find Unknown Lengths of Sides in Similar Triangles

Example 5:

Study the solution for Example 5 part a on page 6.4-18, and record the answer below. Complete part b on your own and check your answer by clicking on the link. If your answer is incorrect, watch the video to find your error.

a. If $\triangle ABC \sim \triangle DEF$, find x.

b. If $\triangle XYZ \sim \triangle PQR$, find n.

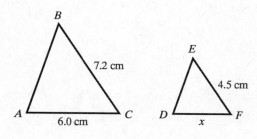
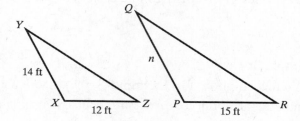

Read and summarize the CAUTION statement on page 6.4-20.

Topic 6.4 Objective 5: Solve Applications Involving Similar Triangles

Example 6:

Complete Example 6 on page 6.4-21 on your own. Check your answer by clicking on the link. If your answer is incorrect, watch the video to find your error.

A support wire is attached to the top of a communication tower and to the ground 180 feet from the bottom of the tower. A worker wants to determine the height of the tower. He holds an 8-foot pole vertically so that its top touches the support wire and its bottom touches the ground, forming similar triangles (see the diagram). He then measures the distance from the bottom of the pole to the point where the support wire is attached to the ground and finds it to be 4.5 feet. Find the height of the tower.

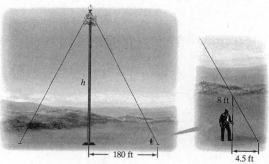

Topic 6.5

Topic 6.5 Guided Notebook

Topic 6.5 Square Roots and the Pythagorean Theorem

Read the list of "THINGS TO KNOW" and review any concepts you are unfamiliar with.

Topic 6.5 Objective 1: Find Square Roots

What does *squaring* a number mean?

What is a **square root**?

Read and summarize the CAUTION statement on 6.5-4.

b is a square root of a, if _____.

How many square roots does a positive number have? Illustrate with an example.

Using your example above, identify the **principal square root** and the **negative square root**.

Read and summarize the CAUTION statement on 6.5-5.

Write the definition of a **square root**.

Topic 6.5

Example 1:
Study the solutions for Example 1 parts a – c on page 6.5-6, and record the answers below.

Evaluate each square root.

a. $\sqrt{25}$ b. $\sqrt{\dfrac{4}{49}}$ c. $\sqrt{0.04}$

Topic 6.5 Objective 2: Approximate Square Roots

What is the definition of a **perfect square**?

Click on the pop-up and list the first 10 whole numbers that are perfect squares.

What is a good way to estimate the square root of a number that is not a perfect square?

Example 2:
Study the solution for Example 2 part a on page 6.5-8. Complete parts b and c on your own and check your answers by clicking on the link. If your answers are incorrect, watch the video to find your error.

For each part, determine the two whole numbers that the square root lies between. Then use a calculator to approximate each square root. Round answers to the nearest thousandth.

a. $\sqrt{57}$ b. $\sqrt{\dfrac{39}{7}}$ c. $\sqrt{19.5}$

Topic 6.5

Summarize the TIP found on page 6.5-10. What method does your calculator use?

Topic 6.5 Objective 3: Use the Pythagorean Theorem

What is a **right triangle**?

Identify the **legs** and **hypotenuse** of the given right triangle.

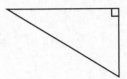

Write down the **Pythagorean Theorem**. Include the diagram and watch the *video animation*.

Example 3:
Study the solution for Example 3 part a on page 6.5-13, and record the answer below. Complete part b on your own and check your answer by clicking on the link. If your answer is incorrect, watch the video to find your error.

Find the length of the missing side for each right triangle.

a.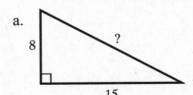

b.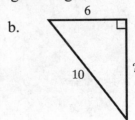

Topic 6.5

Example 4:

Study the solution for Example 4 part a on page 6.5-15, and record the answer below. Complete part b on your own and check your answer by clicking on the link. If your answer is incorrect, watch the video to find your error.

Find the exact length of the missing side of each right triangle. The approximate the length to the nearest tenth.

a.
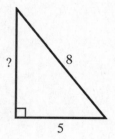

b.
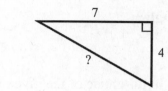

Topic 6.5 Objective 4: Solve Applications Using the Pythagorean Theorem

Example 5:

Study the solution for Example 5 on page 6.5-17, and record the answer below.

Television manufacturers give the screen size of a television by reporting the length of its diagonal. If the length of the screen is 40 inches and its width is 25 inches, approximate the screen size to the nearest tenth of an inch.

Topic 7.1 Guided Notebook

Topic 7.1 Percents, Decimals, and Fractions

Read the list of "THINGS TO KNOW" and review any concepts you are unfamiliar with.

Topic 7.1 Objective 1: Write Percents as Decimals

What does the word **percent** mean?

Write down the equivalent fraction and decimal forms of 87%.

Record the method for **Writing a Percent as a Decimal**.

Summarize the TIP found on page 7.1-4.

Dividing by 100% has two results. What are they?

    1.

    2.

Example 2:
Study the solutions for Example 2 parts a and b on page 7.1-6, and record the answers below. Complete parts c and d on your own and check your answers by clicking on the link. If your answers are incorrect, watch the video to find your error.

Write each percent as a decimal.

  a. 63%   b. 7.25%   c. 248%   d. 0.5%

Topic 7.1

Topic 7.1 Objective 2: Write Decimals as Percents

Record the method for **Writing a Decimal as a Percent**.

Summarize the TIP found on page 7.1-8.

Multiplying by 100% has two results. What are they?

1.

2.

Example 4:
Study the solutions for Example 4 parts a and b on page 7.1-9, and record the answers below. Complete parts c and d on your own and check your answers by clicking on the link. If your answers are incorrect, watch the video to find your error.

Write each decimal as a percent.

a. 0.48　　　　　b. 2.3　　　　　c. 0.0675　　　　　d. 0.009

Read and summarize the CAUTION statement on 7.1-10.

Topic 7.1 Objective 3: Write Percents as Fractions

Record the method for **Writing a Percent as a Fraction or Mixed Number**.

Topic 7.1

When writing a percent as a fraction, dividing by 100% leads to the following three steps:

1.

2.

3.

Summarize the TIP found on page 7.1-12.

Example 6:
Study the solution for Example 6 part a on page 7.1-12, and record the answer below. Complete parts b and c on your own and check your answers by clicking on the link. If your answers are incorrect, watch the video to find your error.

Write each percent as a simplified fraction or mixed number.

a. 8.6% b. 0.25% c. 320%

Example 7:
Study the solution for Example 7 part a on page 7.1-15, and record the answer below. Complete parts b and c on your own and check your answers by clicking on the link. If your answers are incorrect, watch the video to find your error.

Write each percent as a simplified fraction or mixed number.

a. $\frac{3}{8}\%$ b. $4\frac{1}{6}\%$ c. $\frac{20}{3}\%$

Read and summarize the CAUTION statement on 7.1-16.

Copyright © 2014 Pearson Education, Inc.

Topic 7.1

Topic 7.1 Objective 4: Write Fractions as Percents

Record the method for **Writing a Fraction as a Percent**.

Example 8:
Study the solutions for Example 8 parts a and b on page 7.1-17, and record the answers below. Complete parts c and d on your own and check your answers by clicking on the link. If your answers are incorrect, watch the video to find your error.

Write each fraction or mixed number as a percent.

a. $\dfrac{3}{4}$ b. $\dfrac{5}{9}$ c. $\dfrac{19}{16}$ d. $5\dfrac{3}{8}$

Topic 7.1 Objective 5: Perform Conversions Among Percents, Decimals, and Fractions

Example 10:
Study the solutions for Example 10 parts a and b on page 7.1-20, and record the answers below. Complete parts c and d on your own and check your answers by clicking on the link. If your answers are incorrect, watch the video to find your error.

Complete the chart with appropriate percents, decimals, and fractions.

	Percent	Decimal	Fraction
a.	28%		
b.			$\dfrac{17}{40}$
c.		0.875	
d.			$\dfrac{13}{12}$

Topic 7.2

Topic 7.2 Guided Notebook

Topic 7.2 Solving Percent Problems with Equations

Read the list of "THINGS TO KNOW" and review any concepts you are unfamiliar with.

Topic 7.2 Objective 1: Translate Word Statements into Percent Equations

What key word translates to multiplication?

What key word translates to an equal sign?

What are variables, such as x, used to represent?

Write down the general form for a **Percent Equation**.

Complete the table below, found on page 7.2-4.

Quantity	Clue
Percent	
Base	
Amount	

Example 1:
Study the solutions for Example 1 parts a – c on page 7.2-5, and record the answers below.

Translate each word statement into a percent equation. Do not solve.

a. 20% of 190 is what number?

b. What percent of 16 is 40?

c. 65% of what number is 79.3?

145

Topic 7.2

Example 2:
Study the solution for Example 2 part a on page 7.2-6 and record the answer below. Complete parts b and c on your own and check your answers by clicking on the link. If your answers are incorrect, watch the video to find your error.

Translate each word statement into a percent equation. Do not solve.

a. 105 is 140% of what number?

b. 18 is what percent of 34?

c. What number is $7\frac{1}{2}\%$ of 46?

Topic 7.2 Objective 2: Solve Percent Equations

Which property of equality is used in solving a percent equation?

Read and summarize the CAUTION statement on 7.2-8.

Example 3:
Study the solutions for Example 3 parts a and b on page 7.2-8, and record the answers below.

a. What number is 60% of 55?

b. $3\frac{1}{2}\%$ of 90 is what number?

Topic 7.2

Read and summarize the CAUTION statement on 7.2-10.

Example 4:
Study the solution for Example 4 part a on page 7.2-10 and record the answer below. Complete part b on your own and check your answer by clicking on the link. If your answer is incorrect, watch the video to find your error.

a. 85 is what percent of 50?

b. What percent of 140 is 49?

Example 5:
Study the solution for Example 5 part a on page 7.2-12 and record the answer below. Complete part b on your own and check your answer by clicking on the link. If your answer is incorrect, watch the video to find your error.

a. 42% of what number is 63?

b. 14.5 is 20% of what number?

Topic 7.3 Guided Notebook

Topic 7.3 Solving Percent Problems with Proportions

Read the list of "THINGS TO KNOW" and review any concepts you are unfamiliar with.

Topic 7.3 Objective 1: Write Percent Problems as Proportions

What is a proportion?

Write down the general form for a **Percent Proportion**.

Example 1:
Study the solutions for Example 1 parts a and b on page 7.3-5 and record the answers below. Complete part c on your own and check your answer by clicking on the link. If your answer is incorrect, watch the video to find your error.

Translate each word statement into a percent proportion. Do not solve.

a. 12% of 75 is what number?

b. What percent of 50 is 18?

c. 30% of what number is 62.5?

Example 2:
Study the solution for Example 2 part a on page 7.3-7 and record the answer below. Complete parts b and c on your own and check your answers by clicking on the link. If your answers are incorrect, watch the video to find your error.

Translate each word statement into a percent proportion. Do not solve.

a. What number is 9.5% of 77?

b. 230 is 175% of what number?

c. 5 is what percent of 14?

Topic 7.3 Objective 2: <u>Solve Percent Problems Using Proportions</u>

Click on the pop-up box and review how to solve proportions as covered in *Topic 6.2*.

Example 3:
Study the solutions for Example 3 parts a and b on page 7.3-8, and record the answers below.

a. What number is 45% of 80?

b. 7.5% of 140 is what number?

Example 4:
Study the solution for Example 4 part a on page 7.3-11 and record the answer below. Complete part b on your own and check your answer by clicking on the link. If your answer is incorrect, watch the video to find your error.

a. 90 is what percent of 200?

b. What percent of 72 is 93.6?

Read and summarize the CAUTION statement on 7.3-12.

Example 5:
Complete Example 5 parts a and b on page 7.3-14 on your own. Check your answers by clicking on the link. If your answers are incorrect, watch the video to find your error.

a. 52% of what number is 117?

b. 57.5 is 125% of what number?

Topic 7.4 Guided Notebook

Topic 7.4 Applications of Percent

Read the list of "THINGS TO KNOW" and review any concepts you are unfamiliar with.

Topic 7.4 Objective 1: Solve Applications Involving Percents

Example 1:
Study the solution for Example 1 on page 7.4-3, and record the answer below. Include all 6 steps of the problem solving process.

As part of a healthy diet, Amanda keeps her fat consumption to 20% of her daily calories. If she consumes 1875 calories in a day, how many of her calories are from fat?

Step 1:

Step 2:

Step 3:

Step 4:

Step 5:

Step 6:

Summarize the TIP found on page 7.4-5. Then watch the video of the alternate solution.

Topic 7.4

Example 2:
Complete Example 2 on page 7.4-5 on your own. Check your answer by clicking on the link. If your answer is incorrect, watch the video to find your error.

Approximately 54 million U.S. households with a television tuned in to the 2012 Super Bowl between the New York Giants and the New England Patriots. If this was 47% of the total number of U.S. households with televisions, how many U.S. households have a television? Round to the nearest million. (*Source:* The Nielsen Company)

Example 3:
Complete Example 3 on page 7.4-7 on your own. Check your answer by clicking on the link. If your answer is incorrect, watch the video to find your error.

The estimated total cost of attendance for an in-state freshman at Arizona State University was $25,600 for the 2011-2012 academic year. Of this, $9728 was for tuition and fees. What percent of the total cost of attendance was for tuition and fees? (*Source*: https://students.asu.edu)

Topic 7.4

Topic 7.4 Objective 2: Solve Applications Involving a Percent Increase

Percent Increase = $\dfrac{\rule{3cm}{0.4pt}}{\rule{3cm}{0.4pt}}$

Describe how to find the **amount of increase**.

Study the example of the price increase of its bundle plan announced by Netflix™.

Example 4:
Study the solution for Example 4 on page 7.4-11, and record the answer below.

The cost of a 30-second commercial during the 2012 Super Bowl was $3.5 million, up from $3.0 million in 2011. Find the percent of increase. Round to the nearest tenth of a percent. (*Source*: forbes.com)

Read and summarize both of the CAUTION statements on 7.4-12.

Topic 7.4 Objective 3: Solve Applications Involving a Percent Decrease

Percent Decrease = $\dfrac{\rule{3cm}{0.4pt}}{\rule{3cm}{0.4pt}}$

Topic 7.4

Describe how to find the **amount of decrease**.

Read and summarize the CAUTION statement on 7.4-13.

Study the example of the price decrease of its DVD-only plan announced by Netflix™.

Example 5:
Study the solution for Example 5 on page 7.4-15, and record the answer below.

The average number of hours worked per week in the U.S. was 40.2 in 2009 but was only 34.6 in 2011. Find the percent of decrease. Round to the nearest percent. (*Source*: usatoday.com)

Read and summarize the CAUTION statement on 7.4-15.

Topic 7.5 Guided Notebook

Topic 7.5 Percent and Problem Solving: Sales Tax, Commission, and Discount

Read the list of "THINGS TO KNOW" and review any concepts you are unfamiliar with.

Topic 7.5 Objective 1: Compute Sales Tax, Overall Price and Tax Rate

Define **sales tax**.

Write down the **Sales Tax Formula**.

Record the method for **Computing the Overall Price**.

Example 1:
Study the solutions for Example 1 parts a and b on page 7.5-4, and record the answers below.

Peg plans to buy a 3-D television priced at $1299. The tax rate is 7%.

a. How much will Peg have to pay in sales tax?

b. What will be Peg's overall price for the television?

Example 2:
Complete Example 2 on page 7.5-5 on your own. Check your answers by clicking on the link. If your answers are incorrect, watch the video to find your error.

A pair of basketball shoes is priced at $189.79. If the tax rate is 6.5%, find the sales tax and the overall price. Round to the nearest cent.

Topic 7.5

Example 4:
Complete Example 4 on page 7.5-7 on your own. Check your answer by clicking on the link. If your answer is incorrect, watch the video to find your error.

Sam bought a new coat priced at $159.99. When he was paying for the coat, the cashier asked Sam for $170.39. What was the tax rate? Round to the nearest tenth of a percent.

Topic 7.5 Objective 2: Compute Commission and Commission Rate

Define **commission**.

Write down the **Commission Formula**.

Example 5:
Study the solution for Example 5 on page 7.5-9, and record the answer below.

Sherry is a realtor who just sold a house for $235,000. If her commission rate is 1.5% of the selling price, compute Sherry's commission.

Example 6:
Complete Example 6 on page 7.5-10 on your own. Check your answer by clicking on the link. If your answer is incorrect, watch the video to find your error.

A jewelry salesperson earned $300 in commission by selling a wedding ring for $7500. What was the commission rate?

Example 7:

Complete Example 7 on page 7.5-11 on your own. Check your answer by clicking on the link. If your answer is incorrect, watch the video to find your error.

Lindsay, a furniture saleswoman, earns a 2.5% commission rate on her total sales each week. If she earned $338.10 in commission one week, what were Lindsay's sales that week?

Topic 7.5 Objective 3: Compute Discount, Sales Price, and Discount Rate

Define **discount**.

Write down the **Discount Formula**.

Write down the **Sale Price Formula**.

Example 8:

Complete Example 8 on page 7.5-13 on your own. Check your answer by clicking on the link. If your answer is incorrect, watch the video to find your error.

A department store is having a sale on men's clothing. Karl wants to buy a new suit originally priced at $595.89. A sign reads, "Take 35% off the price marked on any suits." Find the discount and the sale price for Karl's suit. Round to the nearest cent.

Topic 7.5

Example 9:
Complete Example 9 on page 7.5-14 on your own. Check your answer by clicking on the link. If your answer is incorrect, watch the video to find your error.

A cordless power saw that normally sells for $125 is on sale for $87.50. Find the discount rate.

Topic 7.6 Guided Notebook

Topic 7.6 Percent and Problem Solving: Interest

Read the list of "THINGS TO KNOW" and review any concepts you are unfamiliar with.

Topic 7.6 Objective 1: Compute Simple Interest

Define **interest**.

Define **principal**.

Define **simple interest**.

Write down the **Simple Interest Formula**.

Summarize the TIP found on page 7.6-3.

Example 1:
Study the solution for Example 1 on page 7.6-4, and record the answer below.

If $2500 is invested for 3 years in a certificate of deposit, or CD, that pays simple interest at an interest rate of 5% per year, how much interest will be earned?

Read and summarize the CAUTION statement on 7.6-4.

Topic 7.6

Summarize the TIP found on page 7.6-5.

Example 2:
Complete Example 2 on page 7.6-5 on your own. Check your answer by clicking on the link. If your answer is incorrect, watch the video to find your error.

Kendall borrowed $1200 for 9 months at a simple interest rate of 8% per year. How much interest will Kendall have to pay?

Record the method for **Finding the Overall Amount for a Loan or Investment**.

Example 3:
Study the solution for Example 3 on page 7.6-7, and record the answer below.

To buy textbooks, a college student borrows $450 at 12% simple interest for 4 months. Find the overall amount that the student must repay at the end of the 4-month period.

Example 4:
Complete Example 4 on page 7.6-8 on your own. Check your answer by clicking on the link. If your answer is incorrect, watch the video to find your error.

If Martha invests $15,000 at a simple interest rate of 6.2% for 1.5 years, what is the overall amount of money that she will have at the end of the 1.5-year period?

Topic 7.6 Objective 2: Compute Compound Interest

What is the difference between simple interest and compound interest?

Watch the *animation* on page 7.6-9 to see how compound interest works.

When investing, which type of interest earns more money? Why?

Interest can be compounded **annually**, which means the interest was computed at the end of the year. Write down the meaning of each of the other types of compound interest.

Semiannually

Quarterly

Monthly

Daily

Topic 7.6

Write down the **Compound Interest Formula**.

Read and summarize the CAUTION statement on 7.6-11.

Record the method for **Finding Interest**.

Example 5:
Study the solution for Example 5 parts a and b on page 7.6-12, and record the answers below.

$3250 is invested at 4% interest compounded semiannually for 3 years.

a. Find the overall amount. Round to the nearest cent.

b. Find the compound interest. Round to the nearest cent.

Example 6:
Complete Example 6 parts a and b on page 7.6-13 on your own. Check your answers by clicking on the link. If your answers are incorrect, watch the video to find your error.

Kejwan put $8500 into a savings account that pays 3.5% compounded quarterly. He plans to grow the money in the account for 5 years.

a. How much money will Kejwan have in the account after 5 years? Round to the nearest cent.

b. How much compound interest will Kejwan have earned over the 5 years? Round to the nearest cent.

Topic 8.1

Topic 8.1 Guided Notebook

Topic 8.1 Lines and Angles

Read the list of "THINGS TO KNOW" and review any concepts you are unfamiliar with.

Topic 8.1 Objective 1: Identify and Name Lines, Segments, Rays, and Angles

Write the definition of a **line**. Draw and label $\overleftrightarrow{AB}$.

Write the definition of a **line segment**. Draw and label $\overline{CD}$.

Read and summarize the first CAUTION statement on 8.1-4.

Write the definition of a **ray**. Draw and label $\overrightarrow{EF}$.

Read and summarize the second CAUTION statement on 8.1-4.

Example 1:
Study the solutions for Example 1 parts a - c on page 8.1-5, and record the answers below.

Identify each figure as a line, segment, or ray. Then use the correct symbol to name the figure.

a. 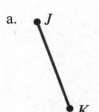   b. 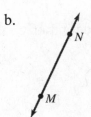   c.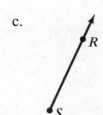

165

Topic 8.1

Write the definition of an **angle**. Draw, label, and identify the parts of ∠ABC.

Read and summarize the CAUTION statement on 8.1-7.

Example 2:
Study the solutions for Example 2 parts a - c on page 8.1-8, and record the answers below.

Using the figure below, name each angle in two other ways.

a. ∠1   b. ∠TQR   c. ∠2

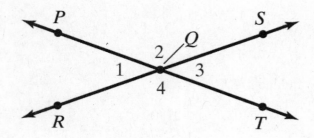

Topic 8.1 Objective 2: Classify Angles as Acute, Obtuse, Right, or Straight

Write the definition of a **straight angle**. Draw and label ∠EFG.

Write the definition of a **right angle**. Draw and label ∠LMN.

Write the definition of an **acute angle**. Write the definition of an **obtuse angle**.

Example 3:
Study the solutions for Example 3 parts a – d on page 8.1-11.

Topic 8.1

Topic 8.1 Objective 3: Identify Complementary and Supplementary Angles and Find Their Measures

Write the definition of **complementary angles**.

Write the definition of **supplementary angles**.

Example 4:
Study the solutions for Example 4 parts a – d on page 8.1-13.

Example 5:
Study the solutions for Example 5 parts a and b on page 8.1-14, and record the answers below. Complete parts c and d on your own and check your answers by clicking on the link. If your answers are incorrect, watch the video to find your error.

Find the measure of each angle described.

a. The complement of an 18° angle.   b. The supplement of a 23° angle.

c. The complement of a 51° angle.   d. The supplement of a 104° angle.

Topic 8.1 Objective 4: Find an Unknown Measure of an Angle Using Given Information

Write the definition of **adjacent angles**. Draw and label a diagram.

Example 6:
Study the solutions for Example 6 parts a and b on page 8.1-15.

Write the definition of a **plane**.

Topic 8.1

Write the definitions of **intersecting** and **parallel lines**.

Write the definition of **vertical angles**.

Example 7:
Study the solutions for Example 7 on page 8.1-18.

Read and summarize the CAUTION statement on 8.1-19.

Example 8:
Study the solutions for Example 8 on page 8.1-20.

Write the definition of a **transversal line**. Copy figure 16 below and identify the **corresponding angles, alternate interior angles**, and **alternate exterior angles**.

Summarize the **Measures of Angles Formed by a Transversal and Parallel Lines**.

Example 9:
Study the solutions for Example 9 on page 8.1-23.

Record the **Sum of the Angles of a Triangle**. Include a diagram.

Topic 8.2

Topic 8.2 Guided Notebook

Topic 8.2 Perimeter, Circumference, and Area

Read the list of "THINGS TO KNOW" and review any concepts you are unfamiliar with.

Topic 8.2 Objective 1: Find the Perimeter of Common Polygons

Write the definition of a **rectangle**. What is the **perimeter formula**?

Read and summarize the CAUTION statement on 8.2-3.

Example 1:
Study the solution for Example 1 on page 8.2-4.

Write the definition of a **square**. What is the **perimeter formula**?

What is a **regular polygon**?

Example 2:
Study the solution for Example 2 on page 8.2-6.

What is a **parallelogram**? What is the **perimeter formula**?

Summarize the TIP found on page 8.2-7.

Example 3:
Study the solution for Example 3 on page 8.2-8.

Topic 8.2

Write the definition of a **trapezoid**. What is the **perimeter formula**?

Example 4:
Study the solution for Example 4 on page 8.2-10.

Write the definition of a **triangle**. What is the **perimeter formula**?

Example 5:
Study the solution for Example 5 on page 8.2-12.

Summarize the TIP found on page 8.2-12.

Example 6:
Complete Example 6 on page 8.2-13 parts a – f on your own. Check your answers by clicking on the link. If your answers are incorrect, watch the video to find your error.

Find the perimeter of each polygon.

a.
9.5 yd

b.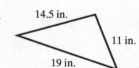
14.5 in.
11 in.
19 in.

c.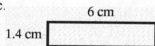
6 cm
1.4 cm

d.
Regular polygon
7 ft

e.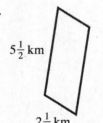
$5\frac{1}{2}$ km
$2\frac{1}{4}$ km

f.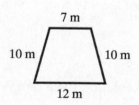
7 m
10 m   10 m
12 m

Topic 8.2

Topic 8.2 Objective 2: Find the Area of Common Polygons

Record the **area formula** for each of the following common polygons:

Rectangle

Triangle

Read and summarize the CAUTION statement on 8.2-15.

Record the **area formula** for each of the following common polygons:

Square

Parallelogram

Trapezoid

Summarize the TIP found on page 8.2-20.

Example 10:
Complete Example 10 on page 8.2-21 parts a – e on your own. Check your answers by clicking on the link. If your answers are incorrect, watch the video to find your error.

Find the area of each polygon.

a.
8.5 km

b.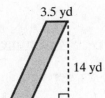
3.5 yd
14 yd

c.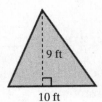
9 ft
10 ft

d.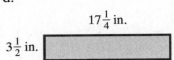
$17\frac{1}{4}$ in.
$3\frac{1}{2}$ in.

e.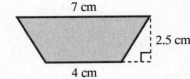
7 cm
2.5 cm
4 cm

171

Topic 8.2

Topic 8.2 Objective 3: Find the Circumference and Area of Circles

Recall and record the formulas for the **Circumference and Area of a Circle**.

Example 11:
Study the solutions for Example 11 part a on page 8.2-23, and record the answers below. Complete part b on your own and check your answers by clicking on the link. If your answers are incorrect, watch the video to find your error.

Find the exact circumference and area of each circle in terms of π. Then approximate the circumference and area using 3.14 for π.

a. 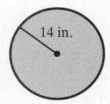   b.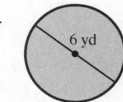

Topic 8.2 Objective 4: Find the Perimeter and Area of Figures Formed from Two or More Common Polygons

Example 13:
Study the solution for Example 13 on page 8.2-27.

Topic 8.2 Objective 5: Solve Applications Involving Perimeter, Circumference, or Area

Review the table of formulas for perimeter, circumference, and area. Write down any that are difficult to remember.

Example 14:
Study the solution for Example 14 on page 8.2-31.

Example 16:
Study the solution for Example 16 on page 8.2-34.

Topic 8.3

Topic 8.3 Guided Notebook

Topic 8.3 Volume and Surface Area

Read the list of "THINGS TO KNOW" and review any concepts you are unfamiliar with.

Topic 8.3 Objective 1: Find the Volume of Common Solids

Write the definition of **plane figures**.

Write the definition of **solids**.

What is the **volume** of a solid?

Watch the video animation on page 8.3-4 to learn about the volume of solids.

Define a **rectangular solid**. Draw and label a diagram, illustrating a **face**, **edge**, and **vertex**.

Record the formula for the **Volume of a Rectangular Solid**.

Example 1:
Study the solution for Example 1 part a on page 8.3-6, and record the answer below. Complete part b on your own and check your answer by clicking on the link. If your answer is incorrect, watch the video to find your error.

Find the volume of each rectangular solid.

a.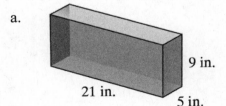

b.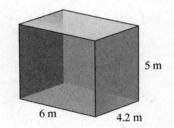

173

Copyright © 2014 Pearson Education, Inc.

Topic 8.3

Define a **cube**.

Record the formula for the **Volume of a Cube**.

Define a **right circular cylinder**.

Record the formula for the **Volume of a Right Circular Cylinder**.

Read and summarize the CAUTION statement on 8.3-10.

Example 3:
Study the solutions for Example 3 parts a and b on page 8.3-11.

Read and summarize the CAUTION statement on 8.3-12.

Define a **sphere**.

Record the formula for the **Volume of a Sphere**.

Example 4:
Study the solutions for Example 4 on page 8.3-13.

Topic 8.3

Define a **right cone** or **cone**.

Record the formula for the **Volume of a Cone**.

Example 5:
Study the solutions for example 5 on page 8.3-16.

Define a **pyramid**. What is a **square pyramid**? A **regular pyramid**?

Example 6:
Study the solution for Example 6 on page 8.3-17.

Topic 8.3 Objective 2: Find the Surface Area of Common Solids

What is the **surface area** of a solid?

Watch the video animation on page 8.3-19 to learn about the surface area of solids.

Record the formula for the **Surface Area of a Rectangular Solid.**

Read and summarize the CAUTION statement on 8.3-21.

Example 7:
Study the solutions for Example 7 parts a and b on page 8.3-22.

Topic 8.3

Record the formula for the **Surface Area of a Cube**.

Study Figure 20 on page 8.3-25 to determine and record the formula for the **Surface Area of a Right Circular Cylinder**.

Example 9:
Study the solution for Example 9 on page 8.3-26.

Record the formula for the **Surface Area of a Sphere**.

Record the formula for the **Surface Area of a Right Cone**.

Record the formula for the **Surface Area of a Regular Pyramid**.

Topic 8.3 Objective 3: Solve Applications Involving Volume or Surface Area

Review the table of formulas for volume and surface area. Write down any that are difficult to remember.

Example 11:
Study the solution for Example 11 on page 8.3-33.

Example 13:
Study the solution for Example 13 on page 8.3-35.

Topic 8.4 Guided Notebook

Topic 8.4 Linear Measurement

Read the list of "THINGS TO KNOW" and review any concepts you are unfamiliar with.

Topic 8.4 Objective 1: Make Conversions Within American Units of Length

List several units of **length** used in the U.S. system of measurement.

What is the definition of a **unit fraction**?

Record the **Unit Fractions for American Units of Length**.

Summarize the second TIP found on page 8.4-5.

Example 2:
Study the solution for Example 2 part a on page 8.4-8, and record the answer below. Complete part b on your own and check your answer by clicking on the link. If your answer is incorrect, watch the video to find your error.

a. Convert 22 feet to yards

b. Convert 2.4 miles to feet

Example 3:
Study the solution for Example 3 on page 8.4-9, and record the answer below.

Convert 78 inches to yards.

Topic 8.4

Topic 8.4 Objective 2: Make Conversions Involving Mixed Units of Length

Example 4:
Study the solution for Example 4 part a on page 8.4-11, and record the answer below. Complete part b on your own and check your answer by clicking on the link. If your answer is incorrect, watch the video to find your error.

a. Convert: 44 feet = _____ yd _____ ft

b. Convert: 3 ft. 9 in. = _____ inches

Topic 8.4 Objective 3: Make Conversions Within Metric Units of Length

Describe the metric system of measurement.

Define **Base Unit**. What is the base unit in the metric system?

Complete the metric system table:

Prefix	kilo	hecto	deka	*base*	deci	centi	milli
Meaning							

Study the **Metric Units of Length** on page 8.4-14.

Complete the following:

Metric Chart for Units of Length

Summarize the TIP found on page 8.4-16.

178

Topic 8.4

Example 5:
Study the solutions for Example 5 parts a and b on page 8.4-17, and record the answers below. Complete parts c and d on your own and check your answers by clicking on the link. If your answers are incorrect, watch the video to find your error.

Perform each of the following conversions.

a. 5 kilometers to meters

b. 142.8 centimeters to meters

c. 650 millimeters to meters

d. 7.3 centimeters to millimeters

Topic 8.4 Objective 4: Make Conversions Between American and Metric Units of Length

What is the key conversion between the American and Metric units of length?

Record the **(Approximate) Unit Fractions for American and Metric Units of Length**.

Example 6:
Study the solution for Example 6 part a on page 8.4-21, and record the answer below. Complete part b on your own and check your answer by clicking on the link. If your answer is incorrect, watch the video to find your error.

a. Convert 13 inches to centimeters.

b. Convert 15 kilometers to miles. Round to the nearest tenth.

Topic 8.4

Example 7:
Study the solution for Example 7 part a on page 8.4-22, and record the answer below. Complete part b on your own and check your answer by clicking on the link. If your answer is incorrect, watch the video to find your error.

a. Convert 7 kilometers to feet. Round to the nearest whole number.

b. Convert 22 meters to inches. Round to the nearest tenth.

Topic 8.5 Guided Notebook

Topic 8.5 Weight and Mass

Read the list of "THINGS TO KNOW" and review any concepts you are unfamiliar with.

Topic 8.5 Objective 1: Make Conversions Within American Units of Weight

Define **weight**.

Define **mass**.

Summarize the TIP found on page 8.5-3.

List several units of **weight** used in the American system of measurement.

Record the **Unit Fractions for American Units of Weight**.

Example 1:
Study the solution for Example 1 on page 8.5-5, and record the answer below.

Convert 9 tons to pounds.

Example 2:
Complete Example 2 on page 8.5-6 on your own. Check your answer by clicking on the link. If your answer is incorrect, watch the video to find your error.

Convert 224 ounces to pounds.

Topic 8.5

Example 3:
Study the solutions for Example 3 parts a and b on page 8.5-6, and record the answers below. Complete parts c and d on your own and check your answers by clicking on the link. If your answers are incorrect, watch the video to find your error.

Make each of the following conversions.

a. 16,500 pounds to tons

b. 4.2 pounds to ounces

c. 54 ounces to pounds

d. $3\frac{1}{8}$ tons to pounds

Topic 8.5 Objective 2: Make Conversions Involving Mixed Units of Weight

Example 4:
Study the solution for Example 4 on page 8.5-8, and record the answer below.

Express 8 lb 10 oz in ounces only.

Example 5:
Study the solution for Example 5 on page 8.5-9, and record the answer below.

Express 195 ounces in mixed units of pounds and ounces.

Topic 8.5 Objective 3: Make Conversions Within Metric Units of Mass

What is the basic unit of mass in the metric system?

Topic 8.5

Study the **Metric Units of Mass** on page 8.5-11. Compare the prefixes to those learned in section 8.4 for the metric units of length.

Summarize the TIP found on page 8.5-12.

Complete the following:

Metric Chart for Units of Mass

Example 6:
Study the solutions for Example 6 parts a and b on page 8.5-13, and record the answers below. Complete parts c and d on your own and check your answers by clicking on the link. If your answers are incorrect, watch the video to find your error.

Make each of the following conversions.

a. 8.5 kilograms to grams

b. 12,300 grams to kilograms

c. 0.2 gram to milligrams

d. 79.25 milligrams to centigrams

Topic 8.5

Topic 8.5 Objective 4: Make Conversions Between American and Metric Units of Weight and Mass

Record the **Approximate Unit Fractions for American and Metric Units of Weight and Mass.**

Summarize the TIP found on page 8.5-16.

Example 7:
Study the solution for Example 7 on page 8.5-17, and record the answer below.

Convert 5.7 pounds to kilograms. Round to the nearest tenth.

Read and summarize the CAUTION statement on 8.5-17.

Example 8:
Complete Example 8 on page 8.5-18 on your own. Check your answer by clicking on the link. If your answer is incorrect, watch the video to find your error.

Convert 12.5 ounces to grams. Round to the nearest tenth.

Topic 8.6 Guided Notebook

Topic 8.6 Capacity

Read the list of "THINGS TO KNOW" and review any concepts you are unfamiliar with.

Topic 8.6 Objective 1: <u>Make Conversions Within American Units of Capacity</u>

Define **capacity**.

List several units of **capacity** used in the American system of measurement.

Record the **Unit Fractions for American Units of Capacity**.

Summarize the TIP found on page 8.6-4.

Example 1:
Study the solution for Example 1 part a on page 8.6-5, and record the answer below. Complete part b on your own and check your answer by clicking on the link. If your answer is incorrect, watch the video to find your error.

Make each indicated conversion.

a. 5 gallons to quarts

b. 96 fluid ounces to cups

Topic 8.6

Example 2:
Study the solution for Example 2 part a on page 8.6-6, and record the answer below. Complete part b on your own and check your answer by clicking on the link. If your answer is incorrect, watch the video to find your error.

Make each indicated conversion.

a. $3\frac{1}{4}$ cups to pints

b. 8.25 quarts to pints

Example 3:
Study the solution for Example 3 on page 8.6-7, and record the answer below.

Convert 3 gallons to cups.

Topic 8.6 Objective 2: Make Conversions Involving Mixed Units of Capacity

Example 4:
Study the solution for Example 4 part a on page 8.6-8, and record the answer below. Complete part b on your own and check your answer by clicking on the link. If your answer is incorrect, watch the video to find your error.

Convert as indicated.

a. 6 gal 3 qt = _____ qt

b. 30 fl oz = _____ c _____ fl oz

Topic 8.6 Objective 3: Make Conversions Within Metric Units of Capacity

What is the basic unit of capacity in the metric system?

Study the **Metric Units of Capacity** on page 8.6-10.

Summarize the TIP found on page 8.6-11.

Complete the following:

Metric Chart for Units of Capacity

Example 5:
Study the solutions for Example 5 parts a and b on page 8.6-13, and record the answers below. Complete parts c and d on your own and check your answers by clicking on the link. If your answers are incorrect, watch the video to find your error.

Make each of the following conversions.

a. 7250 milliliters to liters

b. 24.3 deciliters to milliliters

c. 0.15 liters to milliliters

d. 525 centiliters to liters

Topic 8.6

Topic 8.6 Objective 4: Make Conversions Between American and Metric Units of Capacity

Record the **Approximate Unit Fractions for American and Metric Units of Capacity.**

Example 6:
Study the solution for Example 6 part a on page 8.6-16, and record the answer below. Complete part b on your own and check your answer by clicking on the link. If your answer is incorrect, watch the video to find your error.

Make each indicated conversion. Round each answer to the nearest tenth.

a. 3.5 quarts to liters

b. 16 fluid ounces to milliliters

Topic 8.7 Guided Notebook

Topic 8.7 Time and Temperature

Read the list of "THINGS TO KNOW" and review any concepts you are unfamiliar with.

Topic 8.7 Objective 1: Make Conversions Within Units of Time

List several units of **time** used in the American *and* metric system of measurements.

Record the **Unit Fractions for American Units of Time**.

Example 1:
Study the solutions for Example 1 parts a and b on page 8.7-5, and record the answers below.

a. Convert 3 days to hours.

b. Convert 240 seconds to minutes.

Example 2:
Study the solution for Example 2 part a on page 8.7-7, and record the answer below. Complete part b on your own and check your answer by clicking on the link. If your answer is incorrect, watch the video to find your error.

a. Convert 240 days to weeks.

b. Convert 3.5 hours to minutes.

Topic 8.7

Example 3:
Study the solution for Example 3 on page 8.7-8, and record the answer below.

Convert 3600 minutes to days.

Topic 8.7 Objective 2: Make Conversions Between Fahrenheit and Celsius Temperatures

What are the American units for temperature?

What are the metric units for temperature?

Compare the scales for Celsius and Fahrenheit.

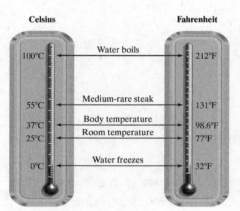

Topic 8.7

Record the **Temperature Conversion Formulas**.

Example 4:
Study the solution for Example 4 part a on page 8.7-12, and record the answer below. Complete part b on your own and check your answer by clicking on the link. If your answer is incorrect, watch the video to find your error.

a. Convert 18°C to Fahrenheit

b. Convert 15°F to Celsius. Round to the nearest degree if necessary.

Topic 8.7

Topic 9.1

Topic 9.1 Guided Notebook

Topic 9.1 Mean, Median, and Mode

Read the list of "THINGS TO KNOW" and review any concepts you are unfamiliar with.

Topic 9.1 Objective 1: Find the Mean

What are the three most common **measures of central tendency**?

What is another term for **mean**? What is the symbol?

Explain how to calculate the mean of a set of data values.

Example 1:
Study the solution to Example 1 on page 9.1-3. Record the answer below.

Valerie keeps track of her gasoline purchases to help her budget monthly expenses. The following data shows the price per gallon (in dollars) for regular unleaded gasoline on her last 10 fill-ups.

    2.97   3.00   3.08   3.10   3.04
    2.96   3.06   3.07   2.95   3.07

Find the mean of the data.

Topic 9.1

Topic 9.1 Objective 2: Find the Median

What is meant by **extreme values**?

Which measure of central tendency is affected by extreme values?

Which measure of central tendency is not greatly affected by extreme values?

Explain how to calculate the median of a set of data values. Be sure to include how to work with an odd-numbered set of data values and an even-numbered set of data values in your explanation.

Example 3:
Study the solutions to Example 3 parts a – d on page 9.1-8. Record the answers below.

The following data show the number of apps downloaded on 8 randomly selected iPads®.

        48 38 43 34 27 22 37 46

a. Find the median.

b. How many values are below the median? How many are above the median?

c. Suppose the value 48 was mistakenly recorded as 84. Find the median using this value.

d. How did changing the value in part c affect the value of the median?

Read and summarize the CAUTION statement on 9.1-10.

Topic 9.1 Objective 3: Find the Mode

What is the **mode** of a set of data values?

Example 4:
Study the solutions to Example 4 parts a – c on page 9.1-11. Record the answers below.

The following data are the snowfall readings (in inches) at Boston's Logan International Airport for Boston's top twelve winter storms. (*Source*: cbsboston.com, February 1, 2011)

$$27.5 \quad 27.1 \quad 26.3 \quad 25.4 \quad 22.5 \quad 21.4$$
$$19.8 \quad 19.4 \quad 18.7 \quad 18.2 \quad 18.2 \quad 18.2$$

a. Find the mode.

b. Suppose the value 27.5 was mistakenly recorded as 72.5. Find the mode using this value.

c. How did changing the value in part b) affect the value of the mode?

Summarize the TIP found on page 9.1-12.

Topic 9.2 Guided Notebook

Topic 9.2 Histograms

Read the list of "THINGS TO KNOW" and review any concepts you are unfamiliar with.

Topic 9.2 Objective 1: Read a Histogram

Define **histogram**.

What is a **class interval**?

What is the **class frequency**?

The histogram in Figure 1 on page 9.2-3 has how many classes? What are they?

What is the **relative frequency** of a histogram?

Example 1:
Study the solutions for Example 1 parts a – c on page 9.2-5, and record the answers below.

The following histogram shows the heights, in inches, for the members of a men's college basketball team. Use the histogram to answer the following questions.

a. Which height range has the most number of players?

b. How many players were between 68 and 71 inches tall?

c. How many players are at least 76 inches tall?

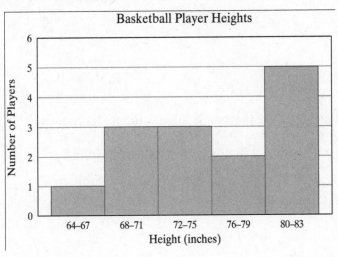

197

Topic 9.2

Example 2:
Complete Example 2 parts a – c on page 9.2-8 on your own. Check your answers by clicking on the link. If your answers are incorrect, watch the video to find your error.

The following histogram shows the speeds, in miles per hour, for a sample of cars along a stretch of highway. Use the histogram to answer the following questions.

a. What percent of cars were driving between 68 and 71 miles per hour?

b. What percent of cars were driving less than 68 miles per hour?

c. Which speed interval had the least number of drivers?

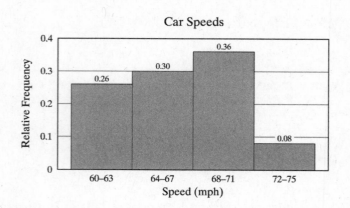

Topic 9.2 Objective 2: Construct a Frequency Table

In order to create a histogram, a **frequency table** must be organized. How is the data organized?

Example 3:
Study the solution for Example 3 on page 9.2-10.

Topic 9.2 Objective 3: Construct a Histogram

To construct a histogram, we first complete _____ or _____ _____, and then we use the table to _____ the bars of the histogram.

Example 4:

Study the solutions for Example 4 on page 9.2-12, and record the answers below.

The following data are the commute distances, in miles, to work for a sample of 30 adults.

5	10	20	35	5	23	14	45	4	8
47	13	11	2	30	12	40	10	16	13
7	4	17	10	20	24	25	30	15	21

Construct a frequency histogram using the given class intervals.

Commute Distance (miles)	Tally	Frequency
1 – 8		
9 – 16		
17 – 24		
25 – 32		
33 – 40		
41 – 48		

Topic 9.2

Example 5:

Complete Example 5 on page 9.2-14 on your own. Check your answers by clicking on the link. If your answers are incorrect, watch the video to find your error.

The following data are the highway gas mileages, in miles per gallon, for a sample of 2012 model year compact cars. (*Source*: fueleconomy.gov)

$$31 \quad 31 \quad 34 \quad 22 \quad 29 \quad 35 \quad 21 \quad 36 \quad 30 \quad 18$$
$$28 \quad 44 \quad 32 \quad 27 \quad 42 \quad 32 \quad 29 \quad 33 \quad 36 \quad 40$$

Construct a relative frequency histogram for the data using the given class intervals.

Gas Mileage (mpg)	Tally	Frequency	Relative Frequency
15-20			
21-26			
27-32			
33-38			
39-44			

Topic 9.3

Topic 9.3 Guided Notebook

Topic 9.3 Counting

Read the list of "THINGS TO KNOW" and review any concepts you are unfamiliar with.

Topic 9.3 Objective 1: Use a Tree Diagram to Count Outcomes

What is a **tree diagram** used for?

Example 1:
Study the solution for Example 1 on page 9.3-4, and record the answer below.

For lunch, Lindsay needs to select a soup and sandwich from the following menu.

Soup	Sandwich
Chicken Noodle	Turkey
Broccoli Cheese	Ham
Vegetable Beef	Roast Beef
	Veggie

Draw the tree diagram to determine the number of possible soup and sandwich combinations.

Example 2:
Complete Example 2 on page 9.3-6 on your own. Check your answer by clicking on the link. If your answer is incorrect, watch the video to find your error.

A customer purchasing an iPad has a choice of color (black or white), storage size (16 GB, 32 GB, or 64 GB), and Internet connection (Wi-Fi only or Wi-Fi + 4G). Draw a tree diagram to determine the number of different iPads that can be purchased.

Topic 9.3 Objective 2: Use the Fundamental Counting Principle to Count Outcomes

Write down the **Fundamental Counting Principle**.

Summarize the TIP found on page 9.3-7.

Example 3:

Study the solution for Example 3 on page 9.3-8, and record the answer below.

Using the information from Figure 3, determine the number of possible sandwiches if you select one meat, one cheese, one topping, and one condiment.

Figure 3

Topic 9.3

Example 4:
Complete Example 4 on page 9.3-9 on your own. Check your answer by clicking on the link. If your answer is incorrect, watch the video to find your error.

A customer ordering a computer online can select between 3 processors (3.60 GHz, 3.50 GHz, 3.10 GHz), 3 hard drive sizes (250 GB, 750 GB, 1 TB), 2 video cards (1 GB, 2 GB), 4 optical drives (DVD player, DVD writer, Blu-ray player, Blu-ray writer), and 2 batteries (6 cell, 9 cell). How many different ways could the customer order a computer?

Example 5:
Complete Example 5 on page 9.3-10 on your own. Check your answer by clicking on the link. If your answer is incorrect, watch the video to find your error.

Use the fundamental counting principle to determine the number of possible five-character passwords with the following conditions:

- the first three characters must be letters
- the letters O and L cannot be used
- the last two characters must be digits
- uppercase and lowercase letters are treated as the same
- no letter may be repeated

Topic 9.3

Topic 9.4 Guided Notebook

Topic 9.4 Probability

Read the list of "THINGS TO KNOW" and review any concepts you are unfamiliar with.

Topic 9.4 Objective 1: Estimate the Probability of an Event

What is the **probability** of an event?

What is an *empirical* probability?

What is a *theoretical* probability?

Define a **probability experiment**. Provide at least two examples.

What are three ways to express a probability?

What does a probability of 0 mean?

What does a probability of 1 mean?

Topic 9.4

Describe how to find the **Estimated Probability of an Event (Empirical Probability)**.

Example 1:
Study the solution for Example 1 on page 9.4-4, and record the answer below.

A basketball player has made 28 of his last 32 free throws. Estimate the probability that he will make his next free throw.

Example 2:
Study the solution for Example 2 part a on page 9.4-5 and record the answer below. Complete part b on your own and check your answer by clicking on the link. If your answer is incorrect, watch the video to find your error.

Mia rolls two six-sided dice 20 times. For each roll, she sums the two values.

6 9 3 4 9 8 4 7 7 4
7 7 10 10 7 6 5 7 7 5

Estimate the probability that rolling two six-sided dice results in

a. a sum of 2.

b. a sum of 9 or more.

Topic 9.4

Topic 9.4 Objective 2: Compute the Probability of an Event for Equally Likely Outcomes

Describe how to find the **Probability of an Event for Equally Likely Outcomes (Theoretical Probability).**

Summarize the TIP found on page 9.4-6.

Example 3:
Study the solution for Example 3 on page 9.4-7, and record the answer below.

A bag of marbles contains 7 yellow marbles, 10 blue marbles, 6 red marbles, and 2 green marbles. What is the probability that a marble selected at random is yellow?

Example 4:
Study the solution for Example 4 part a on page 9.4-8, and record the answer below. Complete part b on your own and check your answer by clicking on the link. If your answer is incorrect, watch the video to find your error.

Find the probability of each of the following events when rolling a pair of six-sided dice.

a. The sum is 2.   b. The sum is 9 or more.

Topic 10.1 Guided Notebook

Topic 10.1 The Real Number System

Read the list of "THINGS TO KNOW" and review any concepts you are unfamiliar with.

Topic 10.1 Objective 1: Classify Real Numbers

Write down the definition for each bold term listed below.

Set   **Element**

Empty set; null set   **Null symbol**

Read and summarize the CAUTION statement on 10.1-3.

Finite sets   **Infinite sets**

Write the definition of **Natural Numbers** and illustrate the set.

Write the definition of **Whole Numbers** and illustrate the set.

Write the definition of **Integers** and illustrate the set.

Write the definition of **Rational Numbers**.

Write the definition of **Irrational Numbers**.

Read and summarize the CAUTION statement on 10.1-7.

Topic 10.1

Write the definition of **Real Numbers**.

Study Figure 1 on page 10.1-8, showing the relationships involving the set of real numbers.

Example 1:
Study the solutions for Example 1 parts a–c on page 10.1-9 and record the answers below. Complete parts d–h on your own and check your answers by clicking on the link. If your answers are incorrect, watch the video to find your error.

Classify each real number as a natural number, whole number, integer, rational number and/or irrational number. Each number may belong to more than one set.

a. 8   b. –4.8   c. $\sqrt{10}$ d. –7

e. $-\dfrac{4}{7}$ f. $\sqrt{25}$ g. 0   h. $3.\overline{45}$

Read and summarize the CAUTION Statement on page 10.1-10.

Topic 10.1 Objective 2: Plot Real Numbers on a Number Line

Draw the **real number line** labeling the following: Negative real numbers, Positive real numbers, Zero and the Origin.

What does it mean to **plot**, or **graph**, a real number?

Example 2:
Study the solutions for Example 2 on page 10.1-12.

Topics 10.1 Objective 3: Find the Opposite of a Real Number

What is the definition of **Opposites**?

Topic 10.1

What is the **Double-Negative Rule**?

What is the procedure for **Finding the Opposite of a Real Number**?

Example 3:
Study the solutions for Example 3 parts a - d on page 10.1-15.

Topic 10.1 Objective 4: Find the Absolute Value of a Real Number

Write down the definition for **Absolute Value.**

Example 4:
Study the solutions for Example 4 parts a and b on page 10.1-17 and record the answers below. Complete parts c–e on your own and check your answers by clicking on the link. If your answers are incorrect, watch the video to find your error.

Find each absolute value.

a. $|3|$ b. $|-5|$ c. $|-1.5|$ d. $\left|-\dfrac{7}{2}\right|$ e. $|0|$

Topic 10.1 Objective 5: Use Inequality Symbols to Order Real Numbers

Record the method used to find the **Order of Real Numbers**.

   1.

   2.

   3.

Example 5:
Complete Example 5 on page 10.1-20 and record the answers below.

Fill in the blank with the correct symbol, <, >, or = to make a true statement.

a. 0 ___ 3   b. −3.7 ___ −1.5   c. $-\dfrac{5}{4}$ ___ −1.25   d. $\dfrac{4}{5}$ ___ $\dfrac{5}{9}$

Topic 10.1

What are **non-strict inequalities**?

What are **strict inequalities**?

Example 6:
Study the solutions for Example 6 parts a and b on pages 10.1-22 and record the answers below. Complete parts c–e on your own and check your answers by clicking on the link. If your answers are incorrect, watch the video to find your error.

Determine if each statement is true or false.

a. $\dfrac{7}{10} \leq 0.7$ b. $-8 \geq 4$ c. $-2 \geq -4$ d. $\dfrac{7}{3} \leq 1.\overline{3}$ e. $-\dfrac{9}{4} \neq -2.75$

Topic 10.1 Objective 6: Translate Word Statements Involving Inequalities

Study Figure 11 on page 10.1-24.

Example 8:
Study the solution for Example 8 part a on page 10.1-26 and record the answer below. Complete parts b and c on your own and check your answers by clicking on the link. If your answers are incorrect, watch the video to find your error.

Use real numbers and write an inequality that represents the given comparison.

a. Orchid Island Gourmet Orange Juice sells for $6, which is more than Florida's Natural Premium Orange Juice that sells for $3.

b. In 2008 there were 4983 identity thefts reported in Colorado, which is different than the 4433 reported in Missouri during the same year. (*Source:* census.gov/compendia/)

c. On May 20, 2010, the Dow Jones Industrial Average closed at 10,068.01 which was less than on May 19, 2010, when it closed at 10,444.37. (*Source:* dowjonesclose.com/)

Topic 10.2 Guided Notebook

Topic 10.2 Adding and Subtracting Real Numbers

Read the list of "THINGS TO KNOW" and review any concepts you are unfamiliar with.

Topic 10.2 Objective 1: Add Two Real Numbers with the Same Sign

What is result of adding numbers called? What are the numbers being added called?

What are the three steps for **Adding Two Real Numbers with the Same Sign**?

1.

2.

3.

Example 1:
Study the solutions for Example 1 on page 10.2-3 and record the answers below. View the popups on page 10.2-4 for the visualization of examples 1a and 1b.

Add
a. $2 + 5$
b. $-4 + (-3)$

Example 2:
Study the solutions for Example 2 parts a and b on page 10.2-4 and record the answers below. Complete parts c and d on your own and check your answers by clicking on the link. If your answers are incorrect, watch the video to find your error.

Add.
a. $-3.65 + (-7.45)$
b. $\dfrac{4}{5} + \dfrac{13}{5}$
c. $-\dfrac{3}{5} + \left(-\dfrac{7}{2}\right)$
d. $-3\dfrac{1}{3} + \left(-5\dfrac{1}{4}\right)$

Topic 10.2 Objective 2: Add Two Real Numbers with Different Signs

What are the three steps for **Adding Two Real Numbers with Different Signs?**

1.

2.

3.

Topic 10.2

Example 3:
Study the solutions for Example 3 on page 10.2-7 and record the answers below. View the popups on page 10.2-9 for the visualization of example 3a and 3b on a number line.

Add.
a. $7 + (-4)$
b. $-6 + 4$
c. $6 + (-6)$

What are **additive inverses**?

What is the rule for **Adding a Real Number and Its Opposite**?

What is the procedure for **Adding Two Real Numbers**?

1.

2.

Example 4:
Study the solutions for Example 4 parts a and b on page 10.2-10 and record the answers below. Complete parts c and d on your own and check your answers by clicking on the link. If your answers are incorrect, watch the video to find your error.

Add.
a. $-12 + (-9)$
b. $7 + (-18)$
c. $\dfrac{4}{3} + \dfrac{5}{6}$
d. $-5.7 + 12.3$

Topic 10.2 Objective 3: Subtract Real Numbers

What is meant by the **difference** of numbers?

What is the **subtrahend**? What is the **minuend**?

What is the definition of **Subtracting Two Real Numbers**?

Example 5:

Study the solutions for Example 5 parts a–d on page 10.2-12 and record the answers below. Complete parts e–g on your own and check your answers by clicking on the link. If your answers are incorrect, watch the video to find your error.

Subtract.

a. $15 - 6$
b. $9 - 17$
c. $-4.5 - 3.2$
d. $\dfrac{7}{3} - \dfrac{2}{5}$

e. $-4.9 - (-2.5)$
f. $7\dfrac{3}{4} - \left(-2\dfrac{1}{5}\right)$
g. $4 - (-4)$

Topic 10.2 Objective 4: Translate Word Statements Involving Addition or Subtraction

Complete the table below found on page 10.2-15.

Key Word	Word Phrase	Mathematical Expression
Sum		
		$4 + 7$
	3 *added to* 8	
		$-4 + 6$

Example 6:

Study the solutions for Example 6 on page 10.2-15 and record the answers below.

Write a mathematical expression for each word phrase.

a. Five more than –8

b. 8.4 increased by 0.17

c. The sum of –4 and –10

d. 15 added to –30

Topic 10.2

Complete the table below found on page 10.2-18.

Key Word	Word Phrase	Mathematical Expression
		9 – 7
Decreased by		
	6 *subtracted from* 3	
		25 – 10

Read and summarize the CAUTION statement on page 10.2-18.

Example 7:
Study the solutions for Example 7 parts a–c on page 10.2-19 and record the answers below. Complete parts d and e on your own and check your answers by clicking on the link. If your answers are incorrect, watch the video to find your error.

Write a **mathematical expression** for each word phrase.

a. Fifteen subtracted from 22.

b. The **difference** of 7 and 12.

c. Eight decreased by 11.

d. 20 less than the difference of 4 and 9.

e. The **sum** of 8 and 13, decreased by 5.

Topic 10.2 Objective 5: Solve Applications Involving Addition or Subtraction of Real Numbers

Example 9:
Work through Example 9 on page 10.2-23 and record the answer below. Check your answer by clicking on the link. If your answer is incorrect, watch the video to find your error.

The record high temperature in Alaska was 100°F recorded in 1915 at Fort Yukon. The record low was –80°F recorded in 1971 at Prospect Creek Camp. What is the difference between these record high and low temperatures? (*Source:* National Climatic Data Center)

Topic 10.3

Topic 10.3 Guided Notebook

Topic 10.3 Multiplying and Dividing Real Numbers

Read the list of "THINGS TO KNOW" and review any concepts you are unfamiliar with.

Topic 10.3 Objective 1: Multiply Real Numbers

Multiplication is simply repeated _____.

What is the **product**? What are **factors**?

Record the steps for **Multiplying Two Real Numbers**.

    1.

    2.

    3.

Example 2:
Study the solution for Example 2 part a on page 10.3-5 and record the answer below. Complete parts b–d on your own and check your answers by clicking on the link. If your answers are incorrect, watch the video to find your error.

Multiply.

a. $\left(-\dfrac{3}{4}\right)\left(-\dfrac{7}{9}\right)$ b. $5 \cdot \dfrac{3}{10}$ c. $\dfrac{3}{8} \times 0$ d. $\left(-\dfrac{2}{3}\right)\left(\dfrac{6}{14}\right)$

Example 3:
Study the solutions for Example 3 parts a and b on page 10.3-6 and record the answers below. Complete parts c and d on your own and check your answers by clicking on your link. If your answers are incorrect, watch the video to find your error.

Multiply.

a. $(1.4)(-3.5)$ b. $(10.32)(0)$ c. $-\dfrac{3}{5} \times 6\dfrac{1}{3}$ d. $(4)(5.8)$

Topic 10.3

Topic 10.3 Objective 2: Divide Real Numbers

In the following equations, identify the **quotient**, **dividend**, and the **divisor.**

$$20 \div 4 = 5 \qquad \frac{20}{4} = 5$$

What is the **reciprocal** or **multiplicative inverse**?

What is the definition of **reciprocals**?

Click on the link for the first CAUTION statement on page 10.3-8, and show why 0 does not have a reciprocal.

Summarize the second CAUTION statement on page 10.3-8.

What is the definition of **Division of Two Real Numbers**?

Record the steps for **Dividing Two Real Numbers Using Absolute Value**.

1.

2.

3.

4.

Example 6:
Study the solutions for Example 6 parts a and b on page 10.3-12 and record the answers below. Complete parts c and d on your own and check your answers by clicking on the link. If your answers are incorrect, watch the video to find your error.

Divide.

a. $\dfrac{48.6}{-3}$ b. $-7\dfrac{2}{5} \div (-3)$ c. $\dfrac{-59.4}{4.5}$ d. $6\dfrac{5}{8} \div 2\dfrac{1}{4}$

According to the CAUTION statement on page 10.3-13, what are the three ways to write a negative quotient?

Topic 10.3 Objective 3: Translate Word Statements Involving Multiplication or Division

Complete the table below that is found on page 10.3-14.

Key Word	Word Phrase	Mathematical Expression
Product		
	−7 times −9	
	One-third of 27	
		0.15(200)
Twice		

What are four symbols that can be used to indicate multiplication?

Example 7:
Study the solutions for Example 7 parts a–d on page 10.3-15 and record the answers below. Complete parts e and f on your own and check your answers by clicking on the link. If your answers are incorrect, watch the video to find your error.

Write a mathematical expression for each word phrase.

a. The product of 3 and −6

b. 30% of 50

c. Three times the sum of 10 and 4

d. Three-fourths of 20, increased by 7

e. The difference of 2 and the product of 8 and 15

f. 3 increased by 15, times 4

Topic 10.3

Rework part 7d as "three-fourths of 20 increased by 7." To see the difference, click on the link on page 10.3-18.

Complete the table below from page 10.3-18.

Key Word	Word Phrase	Mathematical Expression
Quotient		
	24 *divided by* -3	
Per		
		$\dfrac{4}{9}$

Example 8:
Study the solutions for Example 8 parts a and b on page 10.3-19 and record the answers below. Complete parts c and d on your own and check your answers by clicking on the link. If your answers are incorrect, watch the video to find your error.

Write a mathematical expression for each word phrase.

a. The ratio of 10 to 35

b. 60 divided by the sum of 3 and 7

c. The quotient of 20 and 4

d. The difference of 12 and 7, divided by the difference of 8 and -3

Rework part 8d as "the difference of 12 and 7 divided by the difference of 8 and -3." Click on the link on page 10.3-20 to see the difference.

Topic 10.3 Objective 4: Solve Applications Involving Multiplication or Division

Example 10:
Study the solution for Example 10 on page 10.3-22 and record the answer below.

The amount of acid in a solution can be found by multiplying the volume of solution by the percent of the solution (written in decimal form). How much acid is in 20 liters of a 3% solution?

Topic 10.4 Guided Notebook

Topic 10.4 Exponents and Order of Operations

Read the list of "THINGS TO KNOW" and review any concepts you are unfamiliar with.

Topic 10.4 Objective 1: Evaluate Exponential Expressions

Write down an exponential expression. Label and explain the meaning of each of the following in your expression: **exponent, power,** and **base**.

What is the definition of an **exponential expression**?

When we raise a number to the first power, the exponent is _____. Any real number raised to the _____ power is equal to _____.

If no exponent is written it is assumed to be an _____.

Example 1:
Study the solutions for Example 1 on page 10.4-4 and record the answers below.

Evaluate each exponential expression.

a. 4^3 b. $\left(\dfrac{2}{3}\right)^4$ c. $(0.3)^2$

When is a negative sign part of the base? Write an example showing each situation.

Example 2:
Study the solutions for Example 2 parts a and b on page 10.4-6 and record the answers below. Complete part c on your own and check your answer by clicking on the link. If your answer is incorrect, watch the video to find your error.

Evaluate each exponential expression.

a. $(-5)^3$ b. -4^2 c. $(-2)^4$

Topic 10.4

View the link on page 10.4-6 to see how an **exponent** is used to determine the sign when the base of an exponential expression is negative.

Topic 10.4 Objective 2: Use the Order of Operations to Evaluate Numeric Expressions

What is the **Order of Operations**?

1.

2.

3.

4.

Summarize the TIP found on page 10.4-7.

Example 3:
Study the solutions for Example 3 on page 10.4-8 and record the answers below.

Simplify each expression.
a. $10 - 4^2$ b. $12 \div 4 + 8$

Watch the illustrations in the two links found on page 10.4-9. Summarize what you learned.

Example 4:
Complete Example 4 on page 10.4-9 and record the answers below. Check your answers by clicking on the link. If your answers are incorrect, watch the video to find your error.

Simplify each expression.

a. $15 - 3 + 6 - 8 + 7$ b. $-3 \cdot 15 \div 5 \cdot 6 \div 2$

Topic 10.4

Example 5:
Study the solutions for Example 5 on page 10.4-9 and record the answers below.

Simplify each expression.

a. $5+(4-2)^2-3^2$

b. $[5-9]^2+12\div 4$

Example 6:
Complete Example 6 on page 10.4-11 and record the answers below. Check your answers by clicking on the link. If your answers are incorrect, watch the video to find your error.

Simplify each expression.

a. $(-5+8)\cdot 3$

b. $(10-4)^2$

c. $12\div(4+8)$

Write down five **grouping symbols** and their names.

View the **calculator example** shown in the CAUTION statement on page 10.4-11. Make note of the importance of grouping symbols when using a calculator.

We must simplify expressions separately in the _____ and _____ of a fraction before _____.

Read and summarize the CAUTION statement on page 10.4-12.

Topic 10.4

Example 7:
Study the solution for Example 7 part a on page 10.4-12 and record the answer below. Complete part b on your own and check your answer by clicking on the link. If your answer is incorrect, watch the video to find your error.

Simplify each expression.

a. $\dfrac{-2(3)+6^2}{(-4)^2-1}$

b. $|7^2-5(3)|\div 2+8$

What are **nested grouping symbols**?

Example 8:
Study the solution for Example 8 part a on page 10.4-14 and record the answer below. Complete part b on your own and check your answer by clicking on the link. If your answer is incorrect, watch the video to find your error.

Simplify each expression.

a. $\left[2^3-3(5-7)^2\right]\div 6-9$

b. $\dfrac{|-5^2+2^3|-10}{4^2-6\cdot 5}$

Example 9:
Complete Example 9 on page 10.4-15 and record the answers below. Check your answers by clicking on the link. If your answers are incorrect, watch the video to find your error.

Simplify each expression.

a. $\dfrac{3}{10}\cdot\dfrac{5}{2}-\dfrac{1}{2}$

b. $36\div\dfrac{8}{3^2-5}+(-2)^3$

c. $\dfrac{\left|\dfrac{1}{3}-\dfrac{3}{5}\right|}{4}\div\dfrac{1}{2}-1$

224

Topic 10.5

Topic 10.5 Guided Notebook

Topic 10.5 Variables and Properties of Real Numbers

Read the list of "THINGS TO KNOW" and review any concepts you are unfamiliar with.

Topic 10.5 Objective 1: Evaluate Algebraic Expressions

Write down the definitions for the following terms.

Variable

Constant

Algebraic expression

What operation is performed when a constant appears next to a variable?

Describe the process to **Evaluate Algebraic Expressions**.

Read and summarize the CAUTION statement on page 10.5-5.

Example 3:
Study the solution for Example 3 part a on page 10.5-7 and record the answer below. Complete part b on your own and check your answer by clicking on the link. If your answer is incorrect, watch the video to find your error.

Evaluate each algebraic expression for the given values of the variables.

a. $\dfrac{x^2+6}{5x-2}$ for $x=2$
b. $|3y-4|+7y-1$ for $y=-3$

Topic 10.5

Example 4:
Study the solution for Example 4 part a on page 10.5-8 and record the answer below. Complete part b on your own and check your answer by clicking on the link. If your answer is incorrect, watch the video to find your error.

Evaluate each algebraic expression for the given values of the variables.

a. $12a + 7b$ for $a = -4$ and $b = 12$
b. $x^2 - 2xy + 3y^2$ for $x = 3$ and $y = -1$

Topic 10.5 Objective 2: Use the Commutative and Associative Properties

Describe the **Commutative Property of Addition** and show an example.

Describe the **Commutative Property of Multiplication** and show an example.

According to the CAUTION statement on page 10.5-10, what operations **do not** have commutative properties? Why not?

Example 5:
Study the solutions for Example 5 on page 10.5-11 and record the answers below.

Use the given property to rewrite each statement. Do not simplify.

a.   Commutative property of multiplication: $-2(6) = $ _____

b.   Commutative property of addition: $5.03 + 9.2 = $ _____

Describe the **Associative Property of Addition** and show an example.

Describe the **Associative Property of Multiplication** and show an example.

According to the CAUTION statement on page 10.5-12, what operations **do not** have associative properties? Why not?

Topic 10.5

Example 6:
Study the solutions for Example 6 on page 10.5-13 and record the answers below.

Use the given property to rewrite each statement. Do not simplify.

a. Associative property of addition: $\left(\dfrac{2}{3}+\dfrac{1}{6}\right)+\dfrac{5}{6}=$ _____

b. Associative property of multiplication: $5\cdot(2\cdot 13)=$ _____

Example 7:
Study the solutions for Example 7 on page 10.5-14 and record the answers below.

Use the commutative and associative properties to simplify each expression.

a. $(3+x)+7$

b. $(8y)\left(\dfrac{1}{2}\right)$

Topic 10.5 Objective 3: Use the Distributive Property

Describe the **Distributive Property** and show an example.

Show two ways to write the Distributive Property.

Does the Distributive Property apply to subtraction? If yes, show an example.

According to the pop up on page 10.5-17, explain why the Distributive Property extends to more than two terms.

Example 8:
Study the solutions for Example 8 on page 10.5-17 and record the answers below.

Use the distributive property to remove parentheses, and write the product as a sum. Simplify if possible.

a. $9(x+2)$

b. $(7x-5)\cdot 3$

Topic 10.5

How do you find the **Opposite of an Expression**?

Example 9:
Study the solution for Example 9 part a on page 10.5-19 and record the answer below. Complete parts b and c on your own and check your answers by clicking on the link. If your answers are incorrect, watch the video to find your error.

Use the distributive property to remove parentheses, and write the product as a sum. Simplify if possible.

a. $2(4y+3z-5)$
b. $-6(3y-8)$
c. $-(2a-7b+8)$

Example 10:
Study the solutions for Example 10 on page 10.5-20.

Topic 10.5 Objective 4: Use the Identity and Inverse Properties

Show an example of the **Identity Property of Addition**.

Show an example of the **Identity Property of Multiplication**.

Show an example of the **Inverse Property of Addition**.

Show an example of the **Inverse Property of Multiplication**.

Example 11:
Complete Example 11 on page 10.5-23 and record the answers below. Check your answers by clicking on the link. If your answers are incorrect, watch the video to find your error.

Identify the property of real numbers illustrated in each statement.

a. $-4 \cdot 1 = -4$
b. $(-5+5)+x = 0+x$

c. $0+y=y$
d. $\frac{1}{2} \cdot 2x = x$

Topic 10.6 Guided Notebook

Topic 10.6 Simplifying Algebraic Expressions

Read the list of "THINGS TO KNOW" and review any concepts you are unfamiliar with.

Topic 10.6 Objective 1: Identify Terms, Coefficients, and Like Terms of an Algebraic Expression

Write down the definitions for the following.

Terms

Variable terms

Constant terms

Coefficient

Read and summarize the CAUTION statement on page 10.6-4.

Example 1:
Study the solutions Example 1 parts a and b on pages 10.6-4 and record the answers below. Complete part c on your own and check your answer by clicking on the link. If your answer is incorrect, watch the video to find your error.

Determine the number of terms in each expression and list the coefficients for each term.

a. $3x^2 + 7x - 3$ b. $4x^3 - \frac{3}{2}x^2 + x - 1$ c. $3x^2 - 2.3x + x - \frac{3}{4}$

What are **like terms**? Show a pair of like terms and a pair of unlike terms.

Are constants like terms?

Topic 10.6

Example 2:
Study the solution for Example 2 part a on page 10.6-6 and record the answer below. Complete part b on your own and check your answer by clicking on the link. If your answer is incorrect, watch the video to find your error.

Identify the like terms in each algebraic expression.

a. $5x^2 + 3x - 6 + 4x^2 - 7x + 10$

b. $3.5a^2 + 2.1ab + 6.9b^2 - ab + 8a^2$

Topic 10.6 Objective 2: Simplify Algebraic Expressions

Example 3:
Study the solutions for Example 3 parts a and b on pages 10.6-7 and record the answers below. Complete parts c–e on your own and check your answers by clicking on the link. If your answers are incorrect, watch the video to find your error.

Simplify each algebraic expression by combining like terms.

a. $5x - 2x$

b. $6x^2 - 12x - 3x^2 + 4x$

c. $3z - 2z^2 + 7z^2$

d. $6x^2 + 2x + 4x + 3$

e. $-3x + 5 - y + x - 8$

What are the steps for **Simplifying an Algebraic Expression**?
1.

2.

Example 4:
Study the solutions for Example 4 parts a and b on page 10.6-9. Complete parts c and d on your own and check your answers by clicking on the link. If your answers are incorrect, watch the video to find your error.

Simplify each algebraic expression.

c. $5(x - 6) - 3(x - 7)$

d. $2(5z + 1) - (3z - 2)$

Topic 10.6 Objective 3: Write Word Statements as Algebraic Expressions

Write down the six key words that indicate addition.

Write down the six key words that indicate subtraction.

Write down the seven key words that indicate multiplication.

Write down the five key words that indicate division.

What can we use to represent unknown values within a verbal description?

Example 5:
Study the solutions for Example 5 parts a–c on pages 10.6-12 and record the answers below. Complete parts d–f on your own and check your answers by clicking on the link. If your answers are incorrect, watch the video to find your error.

Write each word statement as an **algebraic expression**. Use x to represent the unknown number.

a. Twenty decreased by a number

b. The product of sixteen and a number

c. Five more than twice a number

d. Three-fourths of the square of a number

e. The quotient of 12 and a number, increased by the number

f. The sum of a number and 4, divided by the difference of the number and 9

Topic 10.6

Example 6:
Study the solutions to Example 6 parts a and b on page 10.6-14. Complete part c on your own and check your answer by clicking on the link. If your answer is incorrect, watch the video to find your error.

c. The state of Texas has 10 fewer institutes of higher education than twice the number in Virginia. If we let n = the number of institutes in Virginia, express the number in Texas, in terms of n. (*Source:* Statistical Abstract, 2010)

Topic 10.6 Objective 4: Solve Applied Problems Involving Algebraic Expressions

Example 7:
Study the solutions for Example 7 on page 10.6-15 and record the answers below.

The perimeter of a rectangle is the sum of the lengths of the sides of the rectangle. Use the following rectangle to answer the questions.

a. Write a simplified algebraic expression that represents the perimeter of the rectangle.

b. Use your result from part (a) to find the perimeter if $x = 7$.

Example 8:
Complete Example 8 on page 10.6-17 and record the answers below. Check your answers by clicking on the link. If your answers are incorrect, watch the video to find your error.

Based on data from the National Fire Protection Association, the number of residential property fires, in thousands, is given by $111x^2 - 1366x + 6959$, where x = the number of years after 2000. The number of vehicle fires, in thousands, is given by $-52.5x + 775$. (*Source:* nfpa.org)

a. Write a simplified algebraic expression for the difference between the number of residential property fires and the number of vehicle fires.

b. Use your result from part (a) to estimate the difference in 2010.

Topic 11.1

Topic 11.1 Guided Notebook

Topic 11.1 The Addition and Multiplication Properties of Equality

Read the list of "THINGS TO KNOW" and review any concepts you are unfamiliar with.

Topic 11.1 Objective 1: <u>Identify Linear Equations in One Variable</u>

Write down the definitions for the following terms.
Equation

Algebraic equation

What is the difference between an algebraic expression and an algebraic equation?

Study the interactive video on page 11.1-3 to practice distinguishing between expressions and equations.

What is the definition of a **Linear Equation in One Variable?**

Write down the definitions for the following terms.
First-degree equation

Nonlinear equations

According to the popup on page 11.1-4, what are three of the five examples of nonlinear equations? Why are they nonlinear?

Watch the interactive video on page 11.1-5 and determine which types of equations are linear and which are nonlinear. Copy down any linear equations from the video.

Topic 11.1

Example 1:
Study the solutions for Example 1 on page 11.1-5 and record the answers below.

Determine if each is a linear equation in one variable. If not, state why.
a. $4x + 3 - 2x$
b. $4x + 2 = 3x - 1$

c. $x^2 + 3x = 5$
d. $2x + 3y = 6$

Topic 11.1 Objective 2: Determine If a Given Value Is a Solution to an Equation

Show the steps to show the truth of each of the following equations. View the popup to check.

$-3 + (-2)^2 = -7 + 1$
$|10 - 26| + 4 = 3^2 + 11$

What is the definition of **Solve**?

What is the definition of a **Solution**?

How do you determine if a given value is a solution to an equation?

Read and summarize the CAUTION statement on page 11.1-7.

Example 2:
Study the solutions for Example 2 parts a and b on page 11.1-7. Complete parts c and d on your own and check your answers by clicking on the link. If your answers are incorrect, watch the video to find your error.

Determine if the given value is a solution to the equation.

c. $|a - 6| - 1 = 9 + a^2$; $a = -2$
d. $\frac{3}{5}w - \frac{1}{2} = -\frac{3}{10}w$; $w = \frac{5}{9}$

234

Copyright © 2014 Pearson Education, Inc.

Topic 11.1

Topics 11.1 Objective 3: Solve Linear Equations Using the Addition Property of Equality

What is a **solution set**?

What are **equivalent equations**?

What are the two forms of an **isolated variable**?

What is the **Addition Property of Equality**?

Does this property hold true for subtraction? Why or why not?

Example 3:
Study the solutions for Example 3 on page 11.1-11 and record the answers below. See how to *check* your answers by clicking on the link.

Solve.

a. $x - 5 = 3$

b. $y + \frac{2}{3} = \frac{1}{5}$

Example 5:
Complete Example 5 on page 11.1-14 on your own. Check your answers by clicking on the link. If your answers are incorrect, watch the video to find your error.

Solve.

a. $5x - 3 = 6x + 2$

b. $-5w + 27 = 13 - 4w$

Topic 11.1

Topic 11.1 Objective 4: Solve Linear Equations Using the Multiplication Property of Equality

What is the **Multiplication Property of Equality**?

Does this hold true for division? Why or why not?

According to the CAUTION statement on page 11.1-15, why is the use of zero not allowed?

Example 7:
Complete Example 7 on page 11.1-18 on your own. Check your answers by clicking on the link. If your answers are incorrect, watch the video to find your error.

Solve.

a. $\dfrac{4}{3}x = 52$

b. $2.2x = 6.93$

Topic 11.1 Objective 5: Solve Linear Equations Using Both Properties of Equality

The _____ property is used first for what purpose?

The _____ property is used second for what purpose?

Example 9:
Study the solution to Example 9 part a on page 11.1-21 and record the answer below. Complete part b on your own and check your answer by clicking on the link. If your answer is incorrect, view the video to find your error.

Use the properties of equality to solve each equation.

a. $7y + 4 = 2y - 6$

b. $2x - 14.5 = 0.5x + 50$

Topic 11.2 Guided Notebook

Topic 11.2 Solving Linear Equations in One Variable

Read the list of "THINGS TO KNOW" and review any concepts you are unfamiliar with.

Topic 11.2 Objective 1: Solve Linear Equations Containing Non-Simplified Expressions
What should be done with non-simplified expressions before using the properties of equality?

Example 1:
Study the solution for Example 1 on page 11.2-3 and record the answer below. Check your answer by clicking on the link.

Solve: $4x + 7 - 2x = 5 - 3x - 3$

What should be done if the equation contains grouping symbols?

Example 2:
Complete Example 2 on page 11.2-4 on your own. Check your answer by clicking on the link. If your answer is incorrect, watch the video to find your error.

Solve: $7 - 2(4z - 3) = 3z + 1$

Example 3:
Complete Example 3 on page 11.2-5 on your own. Check your answer by clicking on the link. If your answer is incorrect, watch the video to find your error.

Solve: $2(3x - 1) - 5x = 3 - (3x + 1)$

Topic 11.2

Topic 11.2 Objective 2: Solve Linear Equations Containing Fractions

When an equation contains fractions, how can we make the calculations more manageable?

Example 5:
Complete Example 5 on page 11.2-7 on your own. Check your answer by clicking on the link. If your answer is incorrect, watch the video to find your error.

Solve: $\dfrac{w+3}{2} - 4 = w + \dfrac{1}{3}$

Example 6:
Complete Example 6 on page 11.2-9 on your own. Check your answer by clicking on the link. If your answer is incorrect, watch the video to find your error.

Solve: $\dfrac{5x}{2} - \dfrac{7}{8} = \dfrac{3}{4}x - \dfrac{11}{8}$

Topics 11.2 Objective 3: Solve Linear Equations Containing Decimals; Apply a General Strategy

What is the procedure to remove the decimals in an equation before combining like terms?

Example 7:
Study the solution for Example 7 on page 11.2-10 and record the answer below.

Solve. $1.4x - 3.8 = 6$

Topic 11.2

Example 8:
Complete Example 8 on page 11.2-11 on your own. Check your answer by clicking on the link. If your answer is incorrect, watch the video to find your error.

Solve. $0.1x + 0.03(7 - x) = 0.05(7)$

Read and summarize the CAUTION statement on page 11.2-12.

What are the steps for **A General Strategy for Solving Linear Equations in One Variable**?

1.

2.

3.

4.

5.

6.

Topic 11.2 Objective 4: Identify Contradictions and Identities

What are the three cases for the solution of a linear equation in one variable?

When no variable terms remain and a false statement results the equation is called a(n) _____, which means what?

Topic 11.2

When no variable terms remain and a true statement results the equation is called a(n) _____, which means what?

Example 9:
Study the solution for Example 9 part a on page 11.2-16 and record the answer below. Complete part b on your own and check your answer by clicking on the link. If your answer is incorrect, watch the video to find your error.

Determine if the equation is a contradiction or an identity. State the solution set.

a. $3x + 2(x - 4) = 5x + 7$ b. $3(x - 4) = x + 2(x - 6)$

Topic 11.2 Objective 5: Use Linear Equations to Solve Application Problems

Review the **Strategy for Solving Linear Equations in One Variable.**

Example 10:
Study the solution for Example 10 on page 11.2-17.

Example 11:
Complete Example 11 on page 11.2-18 on your own. Check your answer by clicking on the link. If your answer is incorrect, watch the video to find your error.

In the U.S., the average pounds of red meat eaten, M, is related to the average pounds of poultry eaten, P, by the equation
$$100M = 14000 - 42P$$
Determine the average amount of poultry eaten if the average amount of red meat eaten is 100.1 pounds. (*Source:* U.S. Department of Agriculture)

Topic 11.3 Guided Notebook

Topic 11.3 Introduction to Problem Solving

Read the list of "THINGS TO KNOW" and review any concepts you are unfamiliar with.

Topic 11.3 Objective 1: Translate Sentences into Equations

Complete Table 1 as found on page 11.3-3.

Key Words That Translate to an Equal Sign			

Example 1:
Study the solutions for Example 1 parts a and b on page 11.3-4 and record the answers below. Complete parts c and d on your own and check your answers by clicking on the link. If your answers are incorrect, watch the video to find your error.

Translate each sentence into an equation. Use x to represent each unknown number.

a. Fifty-two less than a number results in -21.

b. Three-fourths of a number, increased by 8, gives the number.

c. The difference of 15 and a number is the same as the sum of the number and 1.

d. If the sum of a number and 4 is multiplied by 2, the result will be 2 less than the product of 4 and the number.

Topic 11.3 Objective 2: Use the Problem-Solving Strategy to Solve Direct Translation Problems

What is a **mathematical model**?

Topic 11.3

What are the six steps of the **Problem-Solving Strategy for Applications of Linear Equations**?

1.

2.

3.

4.

5.

6.

Example 2:
Study the solution for Example 2 on page 11.3-7 and record the answer below.

Five times a number, increased by 17, is the same as 11 subtracted from the number. Find the number.

Example 3:
Complete Example 3 on page 11.3-9 on your own. Check your answer by clicking on the link. If your answer is incorrect, watch the video to find your error.

Four times the difference of twice a number and 5 results in the number increased by 50. Find the number.

Topics 11.3 Objective 3: Solve Problems Involving Related Quantities

For some problems, we need to find _____ quantities that are related in some way.

Example 4:
Study the solution for Example 4 on page 11.3-10.

Example 5:
Complete Example 5 on page 11.3-11 on your own. Check your answer by clicking on the link. If your answer is incorrect, watch the video to find your error.

Disney's *Toy Story* is 11 minutes shorter than its sequel *Toy Story 2*. *Toy Story 3* is 17 minutes longer than *Toy Story 2*. If the total running time for the three movies is 282 minutes, find the running time of each movie. (*Source:* Disney)

Topic 11.3 Objective 4: Solve Problems Involving Consecutive Integers

What are **consecutive integers**?

Give an **Example** of each of the following:

Three consecutive integers

Three consecutive even integers

Three consecutive odd integers

What is the **General Relationship** of each of the following?

Three consecutive integers
x, _____, _____

Three consecutive even integers
x, _____, _____

Three consecutive odd integers
x, _____, _____

Topic 11.3

Read and summarize the CAUTION statement on page 11.3-13.

Example 6:
Complete Example 6 on page 11.3-14 on your own. Check your answer by clicking on the link. If your answer is incorrect, watch the video to find your error.

The sum of two consecutive integers is 79. Find the two integers.

Example 7:
Complete Example 7 on page 11.3-15 on your own. Check your answer by clicking on the link. If your answer is incorrect, watch the video to find your error.

Three consecutive even integers add to 432. Find the three integers.

Topic 11.3 Objective 5: Solve Problems Involving Value

Example 8:
Complete Example 8 on page 11.3-16 on your own. Check your answer by clicking on the link. If your answer is incorrect, watch the video to find your error.

Ethan's cell phone plan costs $34.99 per month for the first 700 minutes, plus $0.35 for each additional minute. If Ethan's bill is $57.39, how many minutes did he use?

Topic 11.4

Topic 11.4 Guided Notebook

Topic 11.4 Formulas

Read the list of "THINGS TO KNOW" and review any concepts you are unfamiliar with.

Topic 11.4 Objective 1: Evaluate a Formula

What is the definition of a **Formula**?

What is the **distance formula**?

Example 1:
Study the solution for Example 1 on page 11.4-3 and record the answer below.

A car travels at an average speed (rate) of 55 miles per hour for 4 hours. How far does the car travel?

Read and summarize the CAUTION statement on page 11.4-4.

Example 2:
Study the solution for Example 2 part a on page 11.4-4 and record the answer below. Complete part b on your own and check your answer by clicking on the link. If your answer is incorrect, watch the video to find your error.

Evaluate the Devine Formula to find the ideal body weight of each person described.
a. A man 72 inches tall

b. A woman 66 inches tall

Look at the two popup boxes on page 11.4-5 and write down the following formulas.
Area

Perimeter

Volume

Topic 11.4

Example 3:
Study the solution for Example 3 part a on page 11.4-6 and record the answer below. Complete part b on your own and check your answer by clicking on the link. If your answer is incorrect, watch the video to find your error.

a. The top of a stainless steel sink is shaped like a square with each side measuring $15\frac{3}{4}$ inches long. How many inches of aluminum molding will be required to surround the outside of the sink?

b. A yield sign has the shape of a triangle with a base of 3 feet and a height of 2.6 feet. Find the area of the sign.

Topic 11.4 Objective 2: Find the Value of a Non-Isolated Variable in a Formula

Example 4:
Study the solution for Example 4 on page 11.4-8 and record the answer below.

The perimeter of a rectangle is given by the formula $P = 2l + 2w$, if $P = 84$ cm and $l = 26$ cm, find w.

What is the formula for **simple interest**? Be sure to identify the variables.

Example 5:
Complete Example 5 on page 11.4-9 on your own. Check your answer by clicking on the link. If your answer is incorrect, watch the video to find your error.

Paige has invested $15,000 in a certificate of deposit (CD) that pays 4% simple interest annually. If she earns $750 in interest when the CD matures, how long has Paige invested the money?

What is the formula that relates Fahrenheit and Celsius measures of temperature?

Topic 11.4

Example 6:
Complete Example 6 on page 11.4-10 on your own. Check your answer by clicking on the link. If your answer is incorrect, watch the video to find your error.

During the month of February, the average high temperature in Montreal, QC is $-4.6°C$ while the average high temperature in Phoenix, AZ is $70.7°F$. (*Source:* World Weather Information Service)

a. What is the equivalent Fahrenheit temperature in Montreal?

b. What is the equivalent Celsius temperature in Phoenix?

Topics 11.4 Objective 3: Solve a Formula for a Given Variable

What does it mean to **solve a formula for a given variable**?

Read and summarize the CAUTION statement on page 11.4-11.

Example 7:
Study the solutions for Example 7 parts a and b on page 11.4-11 and record the answers below. Complete part c on your own and check your answer by clicking on the link. If your answer is incorrect, watch the video to find your error.

Solve each formula for the given variable.

a. Selling price: $S = C + M$ for M

b. Area of a triangle: $A = \frac{1}{2}bh$ for b

c. Perimeter of a rectangle: $P = 2l + 2w$ for l

Read and summarize the CAUTION statement on page 11.4-13

Topic 11.4

Topic 11.4 Objective 4: Use Geometric Formulas to Solve Applications

Example 8:
Study the solution for Example 8 on page 11.4-14 and record the answer below.

An above-ground pool is shaped like a circular cylinder with diameter of 28 ft and a depth of 4.5 ft. If 1 ft^3 ≈ 7.5 gal, how many gallons of water will the pool hold? Use $\pi = 3.14$ and round to the nearest thousand gallons.

Example 9:
Complete Example 9 on page 11.4-15 on your own. Check your answer by clicking on the link. If your answer is incorrect, watch the video to find your error.

Terrence wants to have a new floor installed in his living room, which measures 20 ft by 15 ft. Extending out 3 ft from one wall is a fireplace in the shape of a trapezoid with base lengths of 4 ft and 8 ft.

a. Find the area that needs flooring.

b. If the flooring costs $5.29 per square foot, how much will Terrence pay for the new floor? (Assume there is no wasted flooring.)

Example 10:
Work through Example 10 on page 11.4-17 and record the answer below. Check your answer by clicking on the link. If your answer is incorrect, watch the video to find your error.

The roundabout intersection with the top view shown in the figure will be constructed using concrete pavement 9 inches thick. How many cubic yards of concrete will be needed for the roundabout? Use $\pi \approx 3.14$.

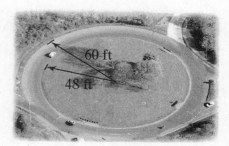

Topic 11.5 Guided Notebook

Topic 11.5 Geometry and Uniform Motion Problem Solving

Read the list of "THINGS TO KNOW" and review any concepts you are unfamiliar with.

Topic 11.5 Objective 1: Solve Problems Involving Geometry Formulas

Example 1:
Study the solution for Example 1 on page 11.5-3 and record the answer below.

A green on a miniature golf course has a rectangular boundary. The length of the boundary is six feet longer than twelve times its width. If the perimeter is 103 feet, what are the dimensions of the green?

Example 2:
Complete Example 2 on page 11.5-5 on your own. Check your answer by clicking on the link. If your answer is incorrect, watch the video to find your error.

The blade of a canoe paddle is in the shape of an isosceles triangle so that two sides have the same length. The two common sides are each 4 inches longer than twice the length of the third side. If the perimeter is 48 inches, find the lengths of the sides of the blade.

Example 3:
Complete Example 3 on page 11.5-6 on your own. Check your answer by clicking on the link. If your answer is incorrect, watch the video to find your error.

The Triangle Drive In has been a local favorite for hamburgers in Fresno, CA since 1963. It is located on a triangular-shaped lot. One side of the lot is 4 meters longer than the shortest side, and the third side is 32 meters less than twice the length of the shortest side. If the perimeter of the lot is 180 meters, find the length of each side.

Topic 11.5

Example 4:
Complete Example 4 on page 11.5-7 on your own. Check your answer by clicking on the link. If your answer is incorrect, watch the video to find your error.

A bathtub is surrounded on three sides by a vinyl wall enclosure. The height of the enclosure is 20 inches less than three times the width, and the length is 5 inches less than twice the width. If the sum of the length, width, and height is 185 inches, what is the volume of the enclosed space?

Topic 11.5 Objective 2: Solve Problems Involving Angles

What are **complementary angles**?

What are **supplementary angles**?

How do you find the following angles?
a. **Complement**

b. **Supplement**

Example 5:
Study the solution for Example 5 on page 11.5-9 and record the answer below.

Find the measure of each complementary angle in the following figure.

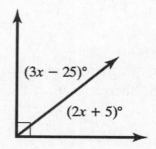

250

Topic 11.5

Example 6:
Complete Example 6 on page 11.5-11 on your own. Check your answer by clicking on the link. If your answer is incorrect, watch the video to find your error.

Find the measure of each supplementary angle in the following figure.

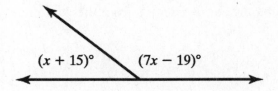

What is the **sum** of the measures of the three angles of a triangle?

Example 7:
Complete Example 7 on page 11.5-13 on your own. Check your answer by clicking on the link. If your answer is incorrect, watch the video to find your error.

Triangle Park in Lexington, KY, has a roughly triangular shape such that the smallest angle measures 10 degrees less than the middle-sized angle. The largest angle measures 30 degrees less than twice the middle-sized angle. Find the measures of all three angles.

Topic 11.5 Objective 3: Solve Problems Involving Uniform Motion

What are **uniform motion** problems?

Topic 11.5

Example 8:
Study the solution for Example 8 on page 11.5-14 and record the answer below.

In January 2010, the U.S government announced plans for the development of high-speed rail projects. A medium-fast passenger train leaves a station traveling 100 mph. Two hours later, a high-speed passenger train leaves the same station traveling 180 mph on a different track. How long will it take the high-speed train to be the same distance from the station as the medium-fast passenger train?

Take notes on the **animation** on page 11.5-16.

Example 9:
Complete Example 9 on page 11.5-16 on your own. Check your answer by clicking on the link. If your answer is incorrect, watch the video to find your error.

Brennan provides in-home healthcare in a rural county and gets reimbursed for mileage. On one particular day he spent 4 hours driving to visit patients. His average speed is 50 mph on the highway but then slows to 30 mph when driving through towns. If he traveled five times as far on the highway as through towns, how far did he travel that day?

Read and summarize the CAUTION statement on page 11.5-18.

Topic 11.6

Topic 11.6 Guided Notebook

Topic 11.6 Percent and Mixture Problem Solving

Read the list of "THINGS TO KNOW" and review any concepts you are unfamiliar with.

Topic 11.6 Objective 1: Solve Problems by Using a Percent Equation

What operation does the word "of" indicate?

What is the **General Equation for Percents**?

How is each of the following identified in percent problems?
  Base

  Percent

  Amount

Read and summarize the CAUTION statement on page 11.6-4.

Example 1:
Study the solution for Example 1 part a on page 11.6-4 and record the answer below. Complete parts b and c on your own and check your answers by clicking on the link. If your answers are incorrect, watch the video to find your error.

Use equations to solve each percent problem.

a. 32 is 40% of what number?

b. 145% of 78 is what number?

c. 8.2 is what percent of 12.5?

Topic 11.6

Example 2:
Study the solution for Example 2 on page 11.6-5 and record the answer below.

6% of a 128 fluid-ounce bottle of bleach is sodium hypochlorite. How many fluid ounces of sodium hypochlorite are in the bottle?

Topic 11.6 Objective 2: Solve Percent Problems Involving Discounts, Markups, and Sales Tax

What are the formulas for **Computing Discounts**?

Example 3:
Study the solution for Example 3 on page 11.6-7 and record the answer below.

A furniture store is going out of business and cuts all prices by 55%. What is the sale price of a sofa with an original price of $1199?

What are the formulas for **Computing Markups**?

Example 4:
Complete Example 4 on page 11.6-8 on your own. Check your answer by clicking on the link. If your answer is incorrect, watch the video to find your error.

A college book store sells all textbooks at a 30% markup over its cost. If the price marked on a biology textbook is $124.28, what was the cost of the book to the store? Round to the nearest cent.

Topic 11.6

What are the formulas for **Computing Sales Tax**?

Example 5:
Study the solution for Example 5 on page 11.6-9 and record the answer below.

Charlotte bought a pair of jeans priced at $51.99. When sales tax was added, she paid an overall price of $55.37. What was the tax rate? Round to the nearest tenth of a percent.

Topics 11.6 Objective 3: Solve Percent of Change Problems

What does **percent of change** describe?

What is a **percent of increase**? What is a **percent of decrease**?

What is the formula for **Percent of Increase**?

Example 6:
Study the solution for Example 6 on page 11.6-11 and record the answer below. See how to *check* your answer by clicking on the link.

Last year, 16,528 students attended City Community College. This year enrollment increased by 3.2%. How many students attend City Community College this year? Round to the nearest whole student.

What is the formula for **Percent of Decrease**?

Topic 11.6

Example 7:
Complete Example 7 on page 11.6-13 on your own. Check your answer by clicking on the link. If your answer is incorrect, watch the video to find your error.

Prior to reorganization in 2010, General Motors (GM) had 91,000 U.S. employees. After the reorganization, GM had 68,500 U.S. employees. By what percent did the number of U.S. employees decrease? (*Source:* General Motors)

Read and summarize the CAUTION statement on page 11.6-13.

Topic 11.6 Objective 4: Solve Mixture Problems

What is the formula for a **Mixture Problem Equation**?

Take notes on the **animation** on page 11.6-14.

Example 9:
Complete Example 9 on page 11.6-17 on your own. Check your answer by clicking on the link. If your answer is incorrect, watch the video to find your error.

How many milliliters of a 25% alcohol solution must be mixed with 10 mL of a 60% alcohol solution to result in a mixture that is 30% alcohol?

Topic 11.7

Topic 11.7 Guided Notebook

Topic 11.7 Linear Inequalities in One Variable

Read the list of "THINGS TO KNOW" and review any concepts you are unfamiliar with.

Topic 11.7 Objective 1: Write the Solution Set of an Inequality in Set-Builder Notation

We use _____ symbols in **inequalities** to show that _____

_____.

Write down an example of **set-builder notation** and explain each part.

Example 1:
Study the solutions for Example 1 parts a - c on page 11.7-4.

Topic 11.7 Objective 2: Graph the Solution Set of an Inequality on a Number Line

In showing solutions on a number line, when is an open circle (∘) used and when is a closed circle (•) used?

Example 2:
Study the solutions for Example 2 parts a and b on page 11.7-5 and record the answer below. Complete parts c–f on your own and check your answers by clicking on the link. If your answers are incorrect, watch the video to find your error.

Graph each solution set on a number line.

a. $\{x \mid x \leq 0\}$ b. $\{x \mid -2 \leq x < 4\}$ c. $\{x \mid x > -1\}$

d. $\{x \mid 3 < x < 7\}$ e. $\{x \mid -1 \leq x \leq 5\}$ f. $\{x \mid x \text{ is any real number}\}$

Topics 11.7 Objective 3: Use Interval Notation to Express the Solution Set of an Inequality

What is the **lower bound** and **upper bound** of an interval?

Topic 11.7

When are parentheses used for the endpoints? When are square brackets used for the endpoints?

Study Table 1 found on page 11.7-8.

Example 3:
Study the solutions for Example 3 parts a and b on page 11.7-9. Complete parts c–f on your own and check your answers by clicking on the link. If your answers are incorrect, watch the video to find your error.

Write each solution set using interval notation.

c. x is less than 4

d. x is between -1 and 5, inclusive

e. $\{x | x \text{ is any real number}\}$

f. $\{x | 8 > x \geq -3\}$

Topic 11.7 Objective 4: Solve Linear Inequalities in One Variable

What is the definition of a **Linear Inequality in One Variable**?

Record the **Addition Property of Inequality**.

Example 4:
Study the solution for Example 4 part a on page 11.7-11. Complete part b on your own and check your answer by clicking on the link. If your answer is incorrect, watch the video to find your error.

Solve each inequality using the addition property of inequality. Write the solution set in interval notation and graph it on a number line.

a. $x + 5 > 4$

b. $y - 3 \leq 1$

What is the **Multiplication Property of Inequality?**

Topic 11.7

Write the **Guidelines for Solving Linear Inequalities in One Variable**

1.

2.

3.

4.

5.

6.

Example 8:
Complete Example 8 on page 11.7-19 on your own. Check your answer by clicking on the link. If your answer is incorrect, watch the video to find your error.

Solve the inequality $4 + 2(3 - x) > 3(2x + 7) + 5$. Write the solution set in set-builder notation, and graph it on a number line.

Example 9:
Complete Example 9 on page 11.7-19 on your own. Check your answer by clicking on the link. If your answer is incorrect, watch the video to find your error.

Solve the inequality $\frac{n}{3} - 4 > -\frac{n}{6} + 1$. Write the solution set in interval notation.

What are a **contradiction** and an **identity**?

Topic 11.7

Example 10:
Study the solution for Example 10 part a on page 11.7-20 and record the answer below. Complete part b on your own. If your answer is incorrect, watch the video to find your error.

Solve the following inequalities. Write each solution set in interval notation.

a. $10 - 2(x+1) > -5x + 3(x+8)$ 	b. $2(5-x) - 2 < 3(x+3) - 5x$

Topic 11.7 Objective 5: Solve Three-Part Inequalities

Example 11:
Study the solution for Example 11 on page 11.7-21.

Read and summarize the CAUTION statement on page 11.7-22.

Topic 11.7 Objective 6: Use Linear Inequalities to Solve Application Problems.

What are the steps in the **Strategy for Solving Application Problems Involving Linear Inequalities**?

1.

2.

3.

4.

5.

6.

Example 13:
Study the solution for Example 13 on page 11.7-24.

Example 14:
Study the solution for Example 13 on page 11.7-25.

Topic 11.8

Topic 11.8 Guided Notebook

Topic 11.8 Compound Inequalities; Absolute Value Equations and Inequalities

Read the list of "THINGS TO KNOW" and review any concepts you are unfamiliar with.

Topic 11.8 Objective 1: Find the Union and Intersection of Two Sets

Write down the definition of **Intersection**. What symbol is used?

Example 2:
Complete Example 2 on page 11.8-4 on your own. Check your answer by clicking on the link. If your answer is incorrect, watch the video to find your error.

Let $A = \{x | x > -2\}$ and $B = \{x | x \leq 5\}$. Find $A \cap B$, the intersection of the two sets.

Write down the definition of **Union**. What symbol is used?

Example 4:
Complete Example 4 on page 11.8-6 on your own. Check your answer by clicking on the link. If your answer is incorrect, watch the video to find your error.

Let $A = \{x | x < -2\}$ and $B = \{x | x \geq 5\}$ Find $A \cup B$, the union of the two sets.

Example 5:
Study the solutions for Example 5 parts a and b on page 11.8-7.

Topic 11.8

Topic 11.8 Objective 2: Solve Compound Linear Inequalities in One Variable

Write down the **Guidelines for Solving Compound Linear Inequalities**.

    1.

    2.

    3.

Example 7:
Complete Example 7 on page 11.8-11 on your own. Check your answer by clicking on the link. If your answer is incorrect, watch the video to find your error.

Solve $9 - 4x < -7$ or $5x + 6 < 3(x+2)$. Graph the solution set and then write it in interval notation.

Explain why the solution to $2x - 3 \leq -1$ and $x - 7 \geq -3$ is a null set.

Explain why the solution to $10x + 7 > 2$ or $3x - 6 \leq 9$ is the set of real numbers.

Topic 11.8 Objective 3: Solve Absolute Value Equations

Write down the **Absolute Value Equation Property**.

Example 9:
Complete Example 9 on page 11.8-16 on your own. Check your answer by clicking on the link. If your answer is incorrect, watch the video to find your error.

Solve. $|1 - 3x| = 4$

Topic 11.8

Read and summarize the CAUTION statement on 11.8-17.

Write down the **Strategy for Solving Absolute Value Equations**.

    1.

    2.

    3.

    4.

Example 13:
Complete Example 13 on page 11.8-20 on your own. Check your answer by clicking on the link. If your answer is incorrect, watch the video to find your error.

Solve. $-3|2-m|+8=2$

Topic 11.8 Objective 4: Solve Absolute Value Inequalities

Study Figures 6 and 7.

Write down the **Absolute Value Inequality Property**.

    1.

    2.

Example 14:
Complete Example 14 on page 11.8-24 on your own. Check your answer by clicking on the link. If your answer is incorrect, watch the video to find your error.

Solve. $|2m-1| \le 5$

Topic 11.8

Read and summarize the CAUTION statement on 11.8-25.

Write down the **Strategy for Solving Absolute Value Inequalities**.

    1.

    2.

    3.

Example 16:
Study the solution for Example 16 on page 11.8-26, and record the answer below.

Solve. $|4x-3|+2 \leq 7$

Example 17:
Complete Example 17 on page 11.8-28 on your own. Check your answer by clicking on the link. If your answer is incorrect, watch the video to find your error.

Solve. $5|1-2x|-3>12$

Topic 12.1

Topic 12.1 Guided Notebook

Topic 12.1 The Rectangular Coordinate System

Read the list of "THINGS TO KNOW" and review any concepts you are unfamiliar with.

Topic 12.1 Objective 1: Read Line Graphs

Why are graphs often used?

What are **line graphs**?

Example 1:
Study the solution for Example 1 parts a and b on page 12.1-4 and record the answers below. Complete parts c–e on your own and check your answers by clicking on the link. If your answers are incorrect, watch the video to find your error.

The following line graph shows the average daily temperature in St. Louis, MO for each month.

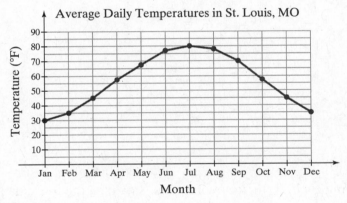

a. What is the average daily temperature in February?

b. What is the average daily temperature in November?

c. In what month is the average daily temperature 70° F?

d. Which month has the highest average daily temperature? What is the average daily temperature for that month?

e. In what months are the average daily temperatures above 65° F?

265

Topic 12.1

Topic 12.1 Objective 2: Identify Points in the Rectangular Coordinate System

Watch the animation on page 12.1-6 and take notes below.

What does the **Cartesian coordinate system** consist of?

Write down the definition of each of the following terms.
perpendicular

***x*-axis**

***y*-axis**

origin

Cartesian plane

quadrants

Draw a picture labeling each of the following: *x*-axis, *y*-axis, the four quadrants, origin

Write down the definition of each of the following terms.
point

ordered pair

Topic 12.1

x-coordinate

y-coordinate

abscissa

ordinate

Example 2:
Complete Example 2 on page 12.1-8 on your own. Check your answers by clicking on the link. If your answers are incorrect, watch the video to find your error.

Use an ordered pair to identify each point on the coordinate plane shown. State the quadrant or axis where each point lies.

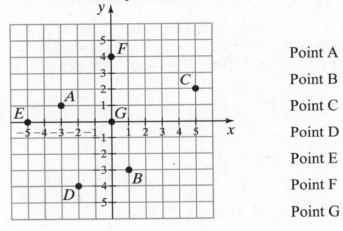

Point A

Point B

Point C

Point D

Point E

Point F

Point G

Read and summarize the CAUTION statement on page 12.1-9.

Topic 12.1 Objective 3: Plot Ordered Pairs in the Rectangular Coordinate System.

We **plot**, or _____, an ordered pair by placing a _____ at its _____ on the coordinate plane.

Example 3:
Study the solutions for Example 3 on page 12.1-10.

Topic 12.1

Topic 12.1 Objective 4: Create Scatter Plots

What can ordered pairs be used to study? Give an example different from the book.

What are **paired data**?

What is a **scatter plot**?

Example 4:
Study the solution for Example 4 on page 12.1-11 and record the answers below.

The table below shows the number of U.S. ethanol plants operating in the month of January for the years 2000–2010. List ordered pairs in the form (*year, number of plants*). Create a scatter plot of the paired data. Do the paired data show a trend? If so, what is the trend?

Year	Number of Plants
2000	54
2001	56
2002	61
2003	68
2004	72
2005	81
2006	95
2007	110
2008	139
2009	170
2010	189

Source: Renewable Fuels Association

Read and summarize the CAUTION statement on page 12.1-12.

Topic 12.2 Guided Notebook

Topic 12.2 Graphing Linear Equations in Two Variables

Read the list of "THINGS TO KNOW" and review any concepts you are unfamiliar with.

Topic 12.2 Objective 1: Determine If an Ordered Pair Is a Solution to an Equation

What is a **Solution to an Equation in Two Variables**?

Example 1:
Study the solutions to Example 1 parts a and b on page 12.2-4 and record the answers below. Complete parts c and d on your own and check your answers by clicking on the link. If your answers are incorrect, watch the video to find your error.

Determine if each ordered pair is a solution to the equation $x + 2y = 8$

a. $(-2, 5)$ b. $(2, 6)$ c. $\left(-11, \dfrac{3}{2}\right)$ d. $(0, 4)$

Topic 12.2 Objective 2: Determine the Unknown Coordinate of an Ordered Pair Solution

What is an **ordered pair solution**?

Example 2:
Study the solution for Example 2 part a on page 12.2-6 and record the answer below. Complete parts b and c on your own and check your answers by clicking on the link. If your answers are incorrect, watch the video to find your error.

Find the unknown coordinate so that each ordered pair satisfies $2x - 3y = 15$.

a. $(6, ?)$ b. $(?, 7)$ c. $\left(-\dfrac{5}{2}, ?\right)$

Topic 12.2 Objective 3: Graph Linear Equation by Plotting Points

What is the **graph of an equation in two variables**?

Topic 12.2

To make such a graph:

1. We can _____ several points that _____ the _____.

2. Then we _____ the points with a _____ or _____.

Write down a **Linear Equation in Two Variables (Standard Form)**.

How many points are required to determine a line?

Example 3:
Study the solution for Example 3 on page 12.2-10.

Example 4:
Study the solution for Example 4 on page 12.2-14.

Example 5:
Complete Example 5 on page 12.2-16 parts a and b on your own. Check your answers by clicking on the link. If your answers are incorrect, watch the video to find your error.

Graph by plotting points.

a. $y = 2x$

b. $3x + 2y = 5$

Topic 12.2 Objective 4: Find x- and y-Intercepts

What are **intercepts**?

What is a **y-intercept**? What is the corresponding ordered pair?

What is an **x-intercept**? What is the corresponding ordered pair?

Topic 12.2

Example 6:
Study the solution for Example 6 on page 12.2-18 and record the answer below.

Find the intercepts of the graph shown in figure 8. What are the *x*-intercepts? What are the *y*-intercepts?

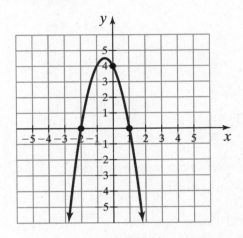

Record the method for **Finding *x*- and *y*-Intercepts of a Graph Given an Equation**.

Example 7:
Study the solution for Example 7 part a on page 12.2-20 and record the answer below. Complete part b on your own and check your answer by clicking on the link. If your answer is incorrect, watch the video to find your error.

Find the *x*- and *y*-intercepts for the graph of each equation.

a. $2x + y = 4$
b. $4x = 3y + 8$

Topic 12.2 Objective 5: Graph Linear Equations Using Intercepts

Study the solution for Example 8 on page 12.2-21 and record the answer below.

Graph $3x - 2y = 6$ using intercepts.

Topic 12.2

Example 9:
Complete Example 9 on page 12.2-23 on your own. Check your answer by clicking on the link. If your answer is incorrect, watch the video to find your error.

Graph $2x = 5y$ using intercepts.

Topic 12.2 Objective 6: Use Linear Equations to Model Data

Example 10:
Complete Example 10 on page 12.2-24 on your own. Check your answers by clicking on the link. If your answers are incorrect, watch the video to find your error.

The number of U.S. drive-in theaters can be modeled by the linear equation $y = -7.5x + 435$, where x is the number of years after 2000. (*Source:* United Drive-In Theater Owners Association, 2009)

a. Sketch the graph of the equation for the year 2000 and beyond.

b. Find the missing coordinate for the ordered pair solution (?, 390).

c. Interpret the point from part (b).

d. Find and interpret the y-intercept.

e. What does the x-intercept represent in this problem?

Topic 12.2 Objective 7: Graph Horizontal and Vertical Lines

What is the graph of the equation $x = a$?

What is the graph of the equation $y = b$?

Example 11:
Study the solution for Example 11 on page 12.2-25.

Topic 12.3

Topic 12.3 Guided Notebook

Topic 12.3 Slope

Read the list of "THINGS TO KNOW" and review any concepts you are unfamiliar with.

Topic 12.3 Objective 1: Find the Slope of a Line Given Two Points

What is a key feature of a line?

How is the **slant** or **steepness** of a line measured?

Write down the definition of **Slope**.

What is the **Slope Formula**? Be sure to include the diagram.

Read and summarize the CAUTION statement on page 12.3-8.

Example 2:
Study the solution for Example 2 on page 12.3-8 and record the answer below.

Find the slope of the line containing the points $(-2, 4)$ and $(1, -3)$.

Summarize the information on **Positive versus Negative Slope** from page 12.3-10.

Example 3:
Study the solution for Example 3 part a on page 12.3-10 and record the answer below. Complete part b on your own and check your answer by clicking on the link. If your answer is incorrect, watch the video to find your error.

Find the slope of the line containing the given points. Simplify if possible.

a. $(-6, -1)$ and $(4, 5)$ b. $(1, 5)$ and $(3, -1)$

Topic 12.3 Objective 2: Find the Slopes of Horizontal and Vertical Lines

Example 4:
Study the solutions for Example 4 parts a and b on page 12.3-11 and record the answers below. Watch the video for detailed solutions.

Find the slope of the line containing the given points. Simplify if possible.

a. $(-3, 2)$ and $(1, 2)$

b. $(4, 2)$ and $(4, -5)$

Summarize the information on **Slopes of Horizontal and Vertical Lines** from page 12.3-13.

Read and summarize the CAUTION statement on page 12.3-13.

Study Figure 18 on page 12.3-13, which summarizes the relationship between the slope and the graph of a linear equation.

Topic 12.3 Objective 3: Graph a Line Using the Slope and a Point

Example 5:
Study the solution for Example 5 on page 12.3-14 and record the answer below.
Graph the line that has slope $m = \frac{3}{2}$ and passes through the point $(1, -2)$.

Example 6:
Complete Example 6 on page 12.3-15 on your own. Check your answer by clicking on the link. If your answer is incorrect, watch the video to find your error.

Graph the line that has slope $m = -3$ and passes through the point $(2, -1)$.

Topic 12.3

Topic 12.3 Objective 4: Find and Use the Slopes of Parallel and Perpendicular Lines

What are **parallel lines**?

Write down the information on **Parallel Lines** found on page 12.3-17.

What are **perpendicular lines**?

Write down the information on **Perpendicular Lines** found on page 12.3-19.

Example 7:
Study the solution for Example 7 on page 12.3-20.

Example 8:
Complete Example 8 on page 12.3-21 on your own. Check your answer by clicking on the link. If your answer is incorrect, watch the video to find your error.

a. Graph a line l_2 that is parallel to l_1 and passes through the point (3,–2)

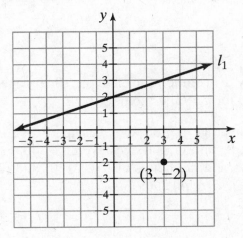

b. Graph a line l_3 that is perpendicular to l_1 and passes through the point (3,–2)

275

Topic 12.3

Topic 12.3 Objective 5: Use Slope in Applications

What is **grade**? Give an example.

What is the slope of a roof called?

Example 9:
Study the solution for Example 9 on page 12.3-22 and record the answer below.

A standard wheelchair ramp should rise no more than 1 foot vertically for every 12 feet horizontally. Find the grade of this ramp. Round to the nearest tenth of a percent. (*Source:* Americans with Disabilities Act Accessibility Guidelines (ADAAG))

What is **average rate of change**? Why is it called this in many applications?

Example 10:
Study the solution for Example 10 on page 12.3-24 and record the answer below. Watch the video for a detailed solution.

The average tuition and fees for U.S. public two-year colleges were $2130 in 1999. The average tuition and fees were $2540 in 2009. Find and interpret the slope of the line connecting the points (1999, 2130) and (2009, 2540). (*Source:* College Board, *Trends in College Pricing 2009*)

Topic 12.4 Guided Notebook
Topic 12.4 Equations of Lines

Read the list of "THINGS TO KNOW" and review any concepts you are unfamiliar with.

Topic 12.4 Objective 1: <u>Determine the Slope and y-Intercept from a Linear Equation</u>

To determine the slope and y-intercept directly from an equation what must be done first?

What is **Slope-Intercept Form**?

Read and summarize the CAUTION statement on page 12.4-4.

Example 2:
Complete Example 2 parts a and b on page 12.4-6 on your own. Check your answers by clicking on the link. If your answers are incorrect, watch the video to find your error.

Find the slope and y-intercept of the given line.

a. $4x - 10y = 0$ b. $y = 4$

Topic 12.4 Objective 2: <u>Use the Slope-Intercept Form to Graph a Linear Equation</u>

Study the animation on page 12.4-7.

Example 4:
Complete Example 4 on page 12.4-9 on your own. Check your answer by clicking on the link. If your answer is incorrect, watch the video to find your error.

Graph the equation $2x + 3y = 9$ using the slope and y-intercept.

Topic 12.4 Objective 3: <u>Write the Equation of a Line Given Its Slope and y-Intercept</u>

If the slope and y-intercept is given, how do you write the equation of the line?

Topic 12.4

Example 5:
Study the solution for Example 5 on page 12.4-10 and record the answers below.

Write an equation of the line with the given slope and *y*-intercept.

a. slope –4; *y*-intercept 3

b. slope $\frac{2}{5}$; *y*-intercept –7

Topic 12.4 Objective 4: Write the Equation of a Line Given Its Slope and a Point on the Line

What is the **Point-Slope Form**?

Example 7:
Complete Example 7 on page 12.4-13 on your own. Check your answer by clicking on the link. If your answer is incorrect, watch the video to find your error.

Use the point-slope form to determine the equation of the line that has slope $-\frac{3}{4}$ and passes through the point (2,–5). Write the equation in slope-intercept form.

Example 8:
Study the solution for Example 8 on page 12.4-15 and record the answers below.

Write the equation of a line that passes through the point (–3,2) and has the given slope.

a. *m* = 0

b. undefined slope

Topic 12.4 Objective 5: Write the Equation of a Line Given Two Points

Example 9:
Complete Example 9 on page 12.4-16 on your own. Check your answer by clicking on the link. If your answer is incorrect, watch the video to find your error.

Write the equation of the line passing through the points (–4,1) and (2,4). Write your answer in slope-intercept form.

Topic 12.4

Complete Table 1 from page 12.4-17

	Slope Average rate of change
	Point-Slope Form Slope is ____ and _____ is a point on the line
$y = mx + b$	
$Ax + By = C$	
	Horizontal Line Slope is _____, and y-intercept is _____
$x = a$	

Topic 12.4 Objective 6: Determine the Relationship Between Two Lines

Parallel lines have the _____ slope but _____ y-intercepts.

Perpendicular lines have _____ slopes.

Coinciding Lines have the _____ slope *and* the _____ y-intercept.

Two lines with different slopes will _____.

Study Table 2 on page 12.4-18.

Example 10:
Study the solution for Example 10 part a on page 12.4-19 and record the answer below. Complete parts b–d on your own and check your answers by clicking on the link. If your answers are incorrect, watch the video to find your error.

For each pair of lines, determine if the lines are parallel, perpendicular, coinciding, or only intersecting.

a. $3y = -2y + 7$
   $3x - 2y = 8$

b. $y = -3x + 1$
   $6x + 2y = 2$

c. $4x - 5y = 15$
   $y = \dfrac{4}{5}x + 1$

d. $3x - 4y = 2$
   $x + 2y = -12$

Topic 12.4

Topic 12.4 Objective 7: Write the Equation of a Line Parallel or Perpendicular to a Given Line

Example 11:
Study the solution for Example 11 part a on page 12.4-20 and record the answer below. Complete part b on your own and check your answer by clicking the link. If your answer is incorrect, watch the video to find your error.

Write the equation of the line that passes through the point (6,−5) and is

a. perpendicular to $6x - 2y = -1$ b. parallel to $y = -2x + 4$

Study Table 3 on page 12.4-22.

Read and summarize the CAUTION statement on page 12.4-22.

Topic 12.4 Objective 8: Use Linear Equations to Solve Applications

Example 12:
Complete Example 12 parts a and b on page 12.4-23 on your own. Check your answers by clicking on the link. If your answers are incorrect, watch the video to find your error.

If attendance at professional football games is 17 million in a given year then the corresponding attendance at college football games is 31 million. Increasing attendance at professional football games to 25 million increases attendance at college football games to 55 million. (*Source:* Statistical Abstract, 2010)

a. Assume that the relationship between professional football attendance (in millions) and college football attendance (in millions) is linear. Find the equation of the line that describes this relationship. Write your answer in slope-intercept form.

b. Use your equation from part (a) to estimate the attendance at college football games if the attendance at professional football games is 21 million

Topic 12.5 Guided Notebook

Topic 12.5 Linear Inequalities in Two Variables

Read the list of "THINGS TO KNOW" and review any concepts you are unfamiliar with.

Topic 12.5 Objective 1: <u>Determine If an Ordered Pair Is a Solution to a Linear Inequality in Two Variables</u>

What is the solution set for a linear inequality in one variable?

Where the solution set of a linear inequality in one variable is typically graphed?

What is the definition of a **Linear Inequality in Two Variables?**

When is an ordered pair a **solution to a linear inequality in two variables**?

Example 1:
Study the solution to Example 1 part a on page 12.5-4 and record the answer below. Complete parts b and c on your own and check your answers by clicking on the link. If your answers are incorrect, watch the video to find your error.

Determine if the given ordered pair is a solution to the inequality $2x - 3y < 6$.

a. $(-1,-2)$ b. $(4,-1)$ c. $(6,2)$

Topic 12.5

Topic 12.5 Objective 2: <u>Graph a Linear Inequality in Two Variables</u>

To find solutions to $x + y < 2$ or $x + y > 2$ what equation do we start with? Graph that equation below. To get more details, click on the link on page 12.5-5.

What is a **half-plane**?

Does the upper half-plane represent the solutions to $x + y < 2$ or $x + y > 2$. Why?

Does the lower half-plane represent the solutions to $x + y < 2$ or $x + y > 2$. Why?

What is a **boundary line**?

What are the **Steps for Graphing Linear Inequalities in Two Variables**?
   1.

   2.

   3.

Read and summarize the **Note** on page 12.5-7

Topic 12.5

Example 2:
Study the solution for Example 2 part a on page 12.5-7 and record the answer below. Complete parts b and c on your own and check your answers by clicking on the link. If your answers are incorrect, watch the video to find your error.

Graph each inequality.

a. $3x - 4y \leq 8$ b. $y > 3x$ c. $y < -2$

Read and summarize the CAUTION statement on 12.5-10.

Topic 12.5 Objective 3: Solve Applications Involving Linear Inequalities in Two Variables

Example 3:
Study the solutions for Example 3 parts a – c on page 12.5-11. Record the answers below.

A piggy bank contains only nickels and dimes with a total value of less than $9. Let n = the number of nickels and d = the number of dimes.

a. Write an inequality describing the possible numbers of coins in the bank.

b. Graph the inequality. Because n and d must be whole numbers, restrict the graph to Quadrant I.

c. Could the piggy bank contain 90 nickels and 60 dimes?

Topic 12.5

Read and summarize the CAUTION statement on page 12.5-12.

Topic 13.1 Guided Notebook

Topic 13.1 Solving Systems of Linear Equations by Graphing

Read the list of "THINGS TO KNOW" and review any concepts you are unfamiliar with.

Topic 13.1 Objective 1: Determine If an Ordered Pair Is a Solution to a System of Linear Equations in Two Variables

What is the definition of a **System of Linear Equations in Two Variables**? Give three examples.

What is the definition of the **Solution to a System of Linear Equations in Two Variables**?

Read and summarize the CAUTION statement on page 13.1-5.

Example 1:
Complete Example 1 parts a and b on page 13.1-5 on your own. Check your answers by clicking on the link. If your answers are incorrect, watch the video to find your error.

Determine if each ordered pair is a solution to the following system:

$$\begin{cases} 2x + 3y = 12 \\ x + 2y = 7 \end{cases}$$

a. $(-3, 6)$ b. $(3, 2)$

Topic 13.1

Topic 13.1 Objective 2: Determine the Number of Solutions to a System Without Graphing

What are the three possible outcomes when two linear equations are graphed? Include a sketch of each.

Write down the definitions for the following terms.

Consistent

Inconsistent

Dependent

Independent

Describe the slopes and y-intercepts of two lines with the following number of solutions.

One solution

No solutions

Infinite number of solutions

Example 2:
Study the solution for Example 2 part a on page 13.1-8 and record the answer below. Complete parts b and c on your own and check your answers by clicking on the link. If your answers are incorrect, watch the video to find your answer.

Determine the number of solutions to each system without graphing.

a. $\begin{cases} y = 3x - 4 \\ 6x + 3y = 8 \end{cases}$
b. $\begin{cases} 2x - 4y = \dfrac{8}{3} \\ 3x - 6y = 4 \end{cases}$
c. $\begin{cases} 5x - 2y = 3 \\ -\dfrac{5}{2}x + y = 7 \end{cases}$

Topics 13.1 Objective 3: Solve Systems of Linear Equations by Graphing

What are the three methods for solving systems of linear equations in two variables?

There is no need to solve which systems by graphing? Why?

What are the steps for **Solving Systems of Linear Equations in Two Variables by Graphing**?

1.

2.

3.

Topic 13.1

Read and summarize the CAUTION statement on page 13.1-11.

Example 3:
Study the solution for Example 3 on page 13.1-12 and record the answer below.

Solve the following system by graphing:
$$\begin{cases} y = 2x+1 \\ y = -x+4 \end{cases}$$

Example 4:
Study the solution for Example 4 on page 13.1-14 and record the answer below.

Solve the following system by graphing:
$$\begin{cases} 3x+y = -2 \\ x+y = 2 \end{cases}$$

Topic 13.2 Guided Notebook

Topic 13.2 Solving Systems of Linear Equations by Substitution

Read the list of "THINGS TO KNOW" and review any concepts you are unfamiliar with.

Topic 13.2 Objective 1: Solve Systems of Linear Equations by Substitution

What are the steps for **Solving Systems of Linear Equations in Two Variables by Substitution**?

1.

2.

3.

4.

Read and summarize the CAUTION statement on page 13.2-4.

Example 1:
Study the solution for Example 1 on page 13.2-5 and record the answer below.

Use the substitution method to solve the following system:

$$\begin{cases} 4x + 2y = 10 \\ y = 3x - 10 \end{cases}$$

Topic 13.2

Example 3:
Complete Example 3 on page 13.2-8 on your own. Check your answer by clicking on the link. If your answer is incorrect, watch the video to find your error.

Solve the following system:

$$\begin{cases} 4x+3y=7 \\ x+9y=-1 \end{cases}$$

Example 4:
Complete Example 4 on page 13.2-9 on your own. Check your answer by clicking on the link. If your answer is incorrect, watch the video to find your error.

Use the substitution method to solve the following system:

$$\begin{cases} 6x-3y=-33 \\ 2x+4y=4 \end{cases}$$

Topic 13.2 Objective 2: Solve Special Systems by Substitution

What are the three possible outcomes when two linear equations are graphed? Include a sketch of each.

Topic 13.2

Describe the slopes, *y*-intercepts, and number of solutions of two lines described below. Also state if they are consistent, inconsistent, dependent, and/or independent. Show an Example for each.

Intersecting Lines

Parallel Lines

Coinciding Lines

When solving algebraically, how are the following situations recognized?

1. The system is independent and consistent.

2. The system is independent and inconsistent.

3. The system is dependent and consistent.

Example 5:
Study the solution for Example 5 on page 13.2-11 and record the answer below.

Use the substitution method to solve the following system:

$$\begin{cases} 2x + 10y = 8 \\ x + 5y = 4 \end{cases}$$

Topic 13.2

Read and summarize the CAUTION statement on page 13.2-15.

Example 6:
Study the solution for Example 6 on page 13.2-15 and record the answer below.

Use the substitution method to solve the following system:

$$\begin{cases} 3x - y = -1 \\ -12x + 4y = 8 \end{cases}$$

Example 7:
Complete Example 7 parts a and b on page 13.2-17 on your own. Check your answers by clicking on the link. If your answers are incorrect, watch the video to find your error.

Use the substitution method to solve the following system:

a. $\begin{cases} \dfrac{1}{4}x + y = 5 \\ x + 4y = 8 \end{cases}$

b. $\begin{cases} -2.4x + 1.5y = -3 \\ 0.8x - 0.5y = 1 \end{cases}$

Topic 13.3 Guided Notebook

Topic 13.3 Solving Systems of Linear Equations by Elimination

Read the list of "THINGS TO KNOW" and review any concepts you are unfamiliar with.

Topic 13.3 Objective 1: Solve Systems of Linear Equations by Elimination

Summarize the **elimination method** as found on page 13.3-3. Include the **Logic for the Elimination Method**.

What is another name for the elimination method?

Example 1:
Study the solution for Example 1 on page 13.3-3 and record the answer below.

Solve the following system.

$$\begin{cases} x + y = 8 \\ x - y = -2 \end{cases}$$

To eliminate a variable, the coefficients of the variable in the two equations must be _____. How can we make this happen?

Topic 13.3

Example 2:
Study the solution for Example 2 on page 13.3-6 record the answer below.

Solve the following system.
$$\begin{cases} x - y = -4 \\ x + 2y = 5 \end{cases}$$

Record the steps for **Solving Systems of Linear Equations in Two Variables by Elimination**.

1.

2.

3.

4.

5.

Read and summarize the CAUTION statement on page 13.3-8.

Topic 13.3

Example 3:
Study the solution for Example 3 on page 13.3-9 record the answer below.

Use the elimination method to solve the following system.

$$\begin{cases} x - 3y = -9 \\ 5x + 4y = -7 \end{cases}$$

Example 5:
Complete Example 5 on page 13.3-12 on your own. Check your answer by clicking on the link. If your answer is incorrect, watch the video to find your error.

Use the elimination method to solve the following system.

$$\begin{cases} 5x - 6y = 20 \\ 4x + 9y = 16 \end{cases}$$

Topic 13.3

Example 6:
Complete Example 6 on page 13.3-13 on your own. Check your answer by clicking on the link. If your answer is incorrect, watch the video to find your error.

Use the elimination method to solve the following system.

$$\begin{cases} x - \dfrac{3}{5}y = \dfrac{4}{5} \\ \dfrac{1}{2}x + 3y = -\dfrac{9}{5} \end{cases}$$

Topic 13.3 Objective 2: Solve Special Systems by Elimination

When solving by elimination, an _____ system will lead to a

_____ and a _____ system will lead to an

_____.

Example 7:
Study the solution for Example 7 part a on page 13.3-14 and record the answer below. Complete part b on your own and check your answer by clicking on the link. If your answer is incorrect, watch the video to find your error.

Use the elimination method to solve each system.

a. $\begin{cases} 3x + y = 6 \\ 6x + 2y = 4 \end{cases}$
b. $\begin{cases} 2x - 8y = 6 \\ 3x - 12y = 9 \end{cases}$

Topic 13.4 Guided Notebook

Topic 13.4 Applications of Linear Systems

Read the list of "THINGS TO KNOW" and review any concepts you are unfamiliar with.

Topic 13.4 Objective 1: Solve Related Quantity Applications Using Systems

What are the six steps for the **Problem-Solving Strategy for Applications Using Systems of Linear Equations**?

1.

2.

3.

4.

5.

6.

Example 1:
Study the solution for Example 1 on page 13.4-4 and record the answer below.

The storage capacity of Deon's external hard drive is 32 times that of his jump drive, a small portable memory device. Together, his two devices have 264 gigabytes of memory. What is the memory size of each device?

Compare the solution process for the same problem from Example 4 Topic 11.3. Use the link found on page 13.4-5.

Topic 13.4

Example 2:
Complete Example 2 on page 13.4-6 on your own. Check your answer by clicking on the link. If your answer is incorrect, watch the video to find your error.

The sum of the ages of Ben and his younger sister Annie is 18 years. The difference of their ages is 4 years. What is the age of each child?

Topic 13.4 Objective 2: Solve Geometry Applications Using Systems

Example 3:
Complete Example 3 on page 13.4-7 on your own. Check your answer by clicking on the link. If your answer is incorrect, watch the video to find your error.

The display panel of a graphing calculator has the shape of a rectangle with a perimeter of 264 millimeters. If the length of the display panel is 18 millimeters longer than the width, find its dimensions.

The measures of two **complementary angles** add to _____, while the measures of two **supplementary angles** add to _____.

Example 4:
Complete Example 4 on page 13.4-8 on your own. Check your answer by clicking on the link. If your answer is incorrect, watch the video to find your error.

Find the measures of two supplementary angles if the measure of the larger angle is 20 degrees less than three times the measure of the smaller angle.

Topic 13.4 Objective 3: Solve Uniform Motion Applications Using Systems

Example 5:
Complete Example 5 on page 13.4-9 on your own. Check your answer by clicking on the link. If your answer is incorrect, watch the video to find your error.

Shawn is training for the Dirty Duo running-and-bicycling race. During a three-hour training session, his total distance cycling and running was 33 miles. If he cycled at a rate of 18 miles per hour and ran at a rate of 6 miles per hour, how much time did he spend doing each activity?

When motions work together the rates are _____, but when they work against each other, the rates are _____.

Example 6:
Complete Example 6 on page 13.4-11 on your own. Check your answer by clicking on the link. If your answer is incorrect, watch the video to find your error.

A jet plane travels 1950 miles in 3.9 hours going with the wind. On the return trip, the plane must fly into the wind and the travel time increases to 5 hours. Find the speed of the jet plane in still air and the speed of the wind. Assume the wind speed is the same for both trips.

Topic 13.4

Topic 13.4 Objective 4: Solve Mixture Applications Using Systems

Example 7:
Complete Example 7 on page 13.4-14 on your own. Check your answer by clicking on the link. If your answer is incorrect, watch the video to find your error.

A shipping company delivered 160 packages one day. The cost of regular delivery is $6.50, and the cost for express delivery is $17.50. Total shipping revenue for the day was $1513. How many of each kind of delivery were made?

Example 8:
Complete Example 8 on page 13.4-16 on your own. Check your answer by clicking on the link. If your answer is incorrect, watch the video to find your error.

A chemist needs eight liters of a 50% alcohol solution but only has a 30% solution and an 80% solution available. How many liters of each solution should be mixed to form the needed solution?

Example 9:
Complete Example 9 on page 13.4-19 on your own. Check your answer by clicking on the link. If your answer is incorrect, watch the video to find your error.

Logan and Payton went to Culver's for lunch. Logan ate two Butterburgers with cheese and a small order of fries for a total of 1801 calories. Payton ate one Butterburger with cheese and two small orders of fries for a total of 1313 calories. How many calories are in a Culver's Butterburger with cheese? How many calories are in a small order of fries?

Topic 13.5 Guided Notebook

Topic 13.5 Systems of Linear Inequalities

Read the list of "THINGS TO KNOW" and review any concepts you are unfamiliar with.

Topic 13.5 Objective 1: Determine If an Ordered Pair Is a Solution to a System of Linear Inequalities in Two Variables

What is the definition of a **System of Linear Inequalities in Two Variables**? Illustrate with two examples.

What is the definition of a **Solution to a System of Linear Inequalities in Two Variables**?

Example 1:
Study the solution for Example 1 part a on page 13.5-4 and record the answer below. Complete parts b and c on your own and check your answers by clicking on the link. If your answers are incorrect, watch the video to find your answer.

Determine if each ordered pair is a solution to the following system of inequalities.

$$\begin{cases} 2x + y \geq -3 \\ x - 4y \leq 12 \end{cases}$$

a. (4, 2) b. (2, –5) c. (0, –3)

Topic 13.5 Objective 2: Graph Systems of Linear Inequalities

What is the **graph of a system of linear inequalities in two variables**?

Topic 13.5

View the animation on page 13.5-6 for an overview of graphing systems of linear inequalities.

Record the **Steps for Graphing Systems of Linear Inequalities**.

   1.

   2.

Read and summarize CAUTION statement on page 13.5-6.

Example 2:
Study the solution for Example 2 on page 13.5-7.

Read and summarize the CAUTION statement on page 13.5-8.

Example 3:
Complete Example 3 on page 13.5-9 on your own. Check your answer by clicking on the link. If your answer is incorrect, watch the video to find your error.

Graph the system of linear inequalities.

$$\begin{cases} x+y<4 \\ x-2y<-2 \end{cases}$$

Read and summarize the CAUTION statement on page 13.5-9.

Topic 13.5

How many solutions does a system of linear inequalities typically have?

What is an **inconsistent system of inequalities**?

Example 4:
Study the solution for Example 4 on page 13.5-10 and record the answer below. Watch the video for a detailed solution.

Graph the system of linear inequalities.

$$\begin{cases} y \leq -\frac{1}{3}x - 3 \\ y > -\frac{1}{3}x + 2 \end{cases}$$

Example 5:
Study the solution for Example 5 on page 13.5-12.

Read and summarize the CAUTION statement on page 13.5-13.

Explain the solution region for the system below by viewing the link on page 13.5-14.

$$\begin{cases} x \geq 0 \\ y \geq 0 \end{cases}$$

Example 6:
Study the solution for Example 6 on page 13.5-14.

Read and summarize the CAUTION statement on page 13.5-15.

Topic 13.5

Topic 13.5 Objective 3: Solve Applications Involving Systems of Linear Inequalities

Example 7:
Study the solution for Example 7 on page 13.5-16 and record your answesr below. Watch the video for a detailed solution.

Savannah is planning a barbeque for her family and friends. She will spend $150 or less to buy hamburger patties that cost $3 per pound and boneless chicken breasts that cost $5 per pound. To limit waste, she will purchase at most 40 pounds of meat all together. Also, the amount of hamburger and chicken purchased must be non-negative. A system of linear inequalities that models this situation is

$$\begin{cases} 3h + 5c \leq 150 \\ h + c \leq 40 \\ h \geq 0 \\ c \geq 0 \end{cases}$$

Where h = pounds of hamburger patties and c = pounds of chicken breasts.

a. Graph the system of linear inequalities.

b. Can Savannah purchase 20 pounds of hamburger patties and 15 pounds of chicken breasts for the barbeque?

c. Can Savannah purchase 10 pounds of hamburger patties and 30 pounds of chicken breasts for the barbeque?

Topic 13.6 Guided Notebook

Topic 13.6 Systems of Linear Equations in Three Variables

Read the list of "THINGS TO KNOW" and review any concepts you are unfamiliar with.

Topic 13.6 Objective 1: Determine If an Ordered Triple Is a Solution to a System of Linear Equations in Three Variables

Write down the definition of a **System of Linear Equations in Three Variables**. Provide one example.

Example 1:
Study the solution for Example 1 part a on page 13.6-4, and record the answer below. Complete part b on your own and check your answer by clicking on the link. If your answer is incorrect watch the video to find your error.

Determine if each ordered triple is a solution to the given system:

$$\begin{cases} 3x + y - 2z = 4 \\ 2x - 2y + 3z = 9 \\ x + y - z = 5 \end{cases}$$

a. $(3, 9, 7)$

b. $(2, -4, -1)$

Topic 13.6 Objective 2: Solve Systems of Linear Equations in Three Variables

Study Figure 16 on page 13.6-7 to learn about the six possibilities for systems of linear equations in three variables.

Topic 13.6

Write down the **Guidelines for Solving a System of Linear Equations in Three Variables by Elimination**.

1.

2.

3.

4.

5.

6.

Example 2:
Study the solution for Example 2 on page 13.6-11.

Example 4:
Complete Example 4 on page 13.6-16 on your own. Check your answer by clicking on the link. If your answer is incorrect, watch the video to find your error.

Solve the following system:

$$\begin{cases} \dfrac{1}{2}x + y + \dfrac{2}{3}z = 2 \\ \dfrac{3}{4}x + \dfrac{5}{2}y - 2z = -7 \\ x + 4y + 2z = 4 \end{cases}$$

Topic 13.6

Example 5:
Study the solution for Example 5 on page 13.6-18, which is an example of an **inconsistent system**. Record the answer below.

Solve the following system:
$$\begin{cases} x - y + 2z = 5 \\ 3x - 3y + 6z = 15 \\ -2x + 2y - 4z = 7 \end{cases}$$

Example 6:
Study the solution for Example 6 on page 13.6-22, which is an example of a **dependent system**. Record the answer below.

Solve the following system:
$$\begin{cases} x - y + 2z = 5 \\ 3x - 3y + 6z = 15 \\ -2x + 2y - 4z = -10 \end{cases}$$

Read and summarize the CAUTION statement on 13.6-24.

Topic 13.6 Objective 3: Use Systems of Linear Equations in Three Variables to Solve Application Problems

Review the six-step **problem-solving strategy** using systems of equations from Topic 13.4.

Example 9:
Complete Example 9 on page 13.6-29 on your own. Check your answer by clicking on the link. If your answer is incorrect, watch the video to find your error.

Wendy ordered 30 T-shirts online for her three children. Small T-shirts cost $4 each, medium T-shirts cost $5 each, and large T-shirts are $6 each. She spent $40 more for the large T-shirts than for the small T-shirts. Wendy's total bill was $154. How many T-shirts of each size did she buy?

Topic 14.1

Topic 14.1 Guided Notebook

Topic 14.1 Exponents

Read the list of "THINGS TO KNOW" and review any concepts you are unfamiliar with.

Topic 14.1 Objective 1: Simplify Exponential Expressions Using the Product Rule

Write down the definitions for the following terms.

Exponential expression

Base

Exponent

What is the **Product Rule for Exponents**?

Example 1:
Study the solutions for Example 1 parts a and b on page 14.1-5 and record the answers below. Complete parts c and d on your own and check your answers by clicking on the link. If your answers are incorrect, watch the video to find your error.

Use the product rule to simplify each expression.

a. $5^4 \cdot 5^6$ b. $x^5 \cdot x^7$ c. $y^3 \cdot y$ d. $b^3 \cdot b^5 \cdot b^4$

Read and summarize the CAUTION statement on page 14.1-5.

Example 2:
Study the solutions for Example 2 parts a and b on page 14.1-6 and record the answers below. Complete part c on your own and check your answer by clicking on the link. If your answer is incorrect, watch the video to find your error.

Simplify using the product rule.

a. $(4x^2)(7x^3)$ b. $(m^4n^2)(m^3n^6)$ c. $(-3a^5b^3)(-8a^2b)$

Topic 14.1

Topic 14.1 Objective 2: Simplify Exponential Expressions Using the Quotient Rule

What is the **Quotient Rule for Exponents**?

Example 3:
Study the solutions for Example 3 parts a and b on page 14.1-8 and record the answers below. Complete parts c and d on your own and check your answers by clicking on the link. If your answers are incorrect, watch the video to find your error.

Use the quotient rule to simplify each expression.

a. $\dfrac{t^9}{t^5}$ b. $\dfrac{7^5}{7^3}$ c. $\dfrac{y^{24}}{y^{15}}$ d. $\dfrac{(-4)^{14}}{(-4)^{11}}$

Example 4:
Study the solutions for Example 4 parts a and b on page 14.1-9 and record the answers below. Complete part c on your own and check your answer by clicking on the link. If your answer is incorrect, watch the video to find your error.

Simplify using the quotient rule.

a. $\dfrac{15x^6}{3x^2}$ b. $\dfrac{a^4 b^9 c^5}{a^2 b^3 c}$ c. $\dfrac{4m^6 n^7}{12m^5 n^2}$

Topic 14.1 Objective 3: Use the Zero-Power Rule

What is the **Zero-Power Rule**?

Example 5:
Study the solutions for Example 5 parts a - e on page 14.1-11 and record the answers below.

Simplify using the zero-power rule.

a. 6^0 b. $(-3)^0$ c. -3^0 d. $(2x)^0$ e. $2x^0$

Topic 14.1 Objective 4: Use the Power-to-Power Rule

What is the **Power-to-Power Rule**?

Example 6:
Study the solution for Example 6 part a on page 14.1-14 and record the answer below. Complete part b on your own and check your answer by clicking on the link. If your answer is incorrect, watch the video to find your error.

Simplify using the power-to-power rule.

a. $(y^5)^6$
b. $[(-2)^3]^5$

Topic 14.1 Objective 5: Use the Product-to-Power Rule

What is the **Product-to-Power Rule**?

Example 7:
Study the solutions for Example 7 parts a and b on page 14.1-16 and record the answers below. Complete parts c and d on your own and check your answers by clicking on the link. If your answers are incorrect, watch the video to find your error.

Simplify using the product-to-product rule.

a. $(mn)^8$
b. $(x^2y)^5$
c. $(3y)^4$
d. $(-4p^5q^3)^2$

Topic 14.1 Objective 6: Use the Quotient-to-Power Rule

What is the **Quotient-to-Power Rule**?

Topic 14.1

Example 8:
Study the solutions for Example 8 parts a and b on page 14.1-18 and record the answers below. Complete parts c and d on your own and check your answers by clicking on the link. If your answers are incorrect, watch the video to find your error.

Simplify using the quotient-to-power rule.

a. $\left(\dfrac{m}{n}\right)^9$
b. $\left(\dfrac{x^2}{y^5}\right)^4$
c. $\left(\dfrac{x}{2}\right)^5$
d. $\left(\dfrac{3x^2}{5y^4}\right)^3$

Topic 14.1 Objective 7: Simplify Exponential Expressions Using a Combination of Rules

Study the four conditions for an exponential expression to be considered **simplified**.

Write down the **Rules for Exponents**.

    Product Rule

    Quotient Rule

    Zero-Power Rule

    Power-to-Power Rule

    Product-to-Power Rule

    Quotient-to-Power Rule

Example 9:
Study the solutions for Example 9 parts a and b on page 14.1-20 and record the answers below. Complete parts c and d on your own and check your answers by clicking on the link. If your answers are incorrect, watch the video to find your error.

Simplify using the rules for exponents.

a. $(c^3)^5(c^2)^6$
b. $\left(\dfrac{15x^8 y^5}{3x^6 y}\right)^2$
c. $(-2w^3 z^2)(-2wz^2)^4$
d. $\dfrac{(4m^2 n^0)(2n^3)^2}{8mn^5}$

Topic 14.2 Guided Notebook
Topic 14.2 Introduction to Polynomials

Read the list of "THINGS TO KNOW" and review any concepts you are unfamiliar with.

Topic 14.2 Objective 1: <u>Classify Polynomials</u>

What is a **simplified term?**

What is the definition of a **Monomial**?

Which of the terms from the interactive video on page 14.2-3 is a monomial? Why?

Write down the definitions for the following terms.
Polynomial

Terms of the polynomial

Simplified polynomial

Polynomials in one variable

How many terms do each of the following have? Give an example of each.
Monomial

Binomial

Trinomial

Polynomial

Topic 14.2

Example 1:
Complete Example 1 parts a – d on page 14.2-5 on your own. Check your answers by clicking on the link. If your answers are incorrect, watch the video to find your error.

Classify each polynomial as a monomial, binomial, trinomial, or none of these.

a. $5x - 7$
b. $\frac{1}{3}x^2$
c. $5x^3 - 7x^2 + 4x + 1$
d. $-2x^3 - 5x^2 + 8x$

Topic 14.2 Objective 2: Determine the Degree and Coefficient of a Monomial

What is the definition of the **Degree of a Monomial**?

Read and summarize the CAUTION statement on page 14.2-6.

What is the definition of the **Coefficient of a Monomial**?

Example 2:
Study the solutions for Example 2 parts a–c on page 14.2-6 and record the answers below. Complete parts d–f on your own and check your answers by clicking on the link. If your answers are incorrect, watch the video to find your error.

Determine the coefficient and degree of each monomial.

a. $4.6x^3$
b. $7x$
c. x^2y^4
d. 12
e. $\frac{3}{4}x^2yz^3$
f. $-2xyz^7$

Topic 14.2 Objective 3: Determine the Degree and Leading Coefficient of a Polynomial

What is the definition of the **Degree of a Polynomial**?

What does it mean for a polynomial to be written in **descending order**?

Topic 14.2

What is **standard form** for polynomials?

Record the definition of the **Leading Coefficient of a Polynomial in One Variable**.

Example 3:
Study the solution for Example 3 part a on page 14.2-9 and record the answer below. Complete part b on your own and check your answer by clicking on the link. If your answer is incorrect, watch the video to find your error.

Write each polynomial in standard form. Then find its degree and leading coefficient.

a. $4.2m - 3m^2 + 1.8 - 7m^3$

b. $\frac{2}{3}x^3 - 3x^2 + 5 - x^4 + \frac{1}{4}x$

Topic 14.2 Objective 4: Evaluate a Polynomial for a Given Value

Example 4:
Study the solutions for Example 4 parts a and b on page 14.2-10 and record the answers below. Complete parts c and d on your own and check your answers by clicking on the link. If your answers are incorrect, watch the video to find your error.

Evaluate the polynomial $x^3 + 3x^2 + 4x - 5$ for the given values of x.

a. $x = -2$ b. $x = 0$ c. $x = 2$ d. $x = \frac{5}{2}$

Topic 14.2 Objective 5: Simplify Polynomials by Combining Like Terms

How do you simplify a polynomial?

Topic 14.2

Example 5:
Study the solutions for Example 5 parts a and b on page 14.2-12 and record the answers below. Complete parts c–e on your own and check your answers by clicking on the link. If your answers are incorrect, watch the video to find your error.

Simplify each polynomial by combining like terms.

a. $3x^2 + 8x - 4x + 2$

b. $2.3x - 3 - 5x + 8.4$

c. $2x + 3x^2 - 6 + x^2 - 2x + 9$

d. $\frac{2}{3}x^2 + \frac{1}{5}x - \frac{1}{10}x - \frac{1}{6}x^2 + \frac{1}{4}$

e. $6x^3 + x^2 - 7$

Topic 14.3

Topic 14.3 Guided Notebook

Topic 14.3 Adding and Subtracting Polynomials

Read the list of "THINGS TO KNOW" and review any concepts you are unfamiliar with.

Topic 14.3 Objective 1: Add Polynomials

Review how to simplify algebraic expressions.

What is the procedure for **Adding Polynomials**?

Example 1:
Study the solution for Example 1 on page 14.3-3 and record the answer below.

Add $(2x + 8) + (7x - 3)$.

Read and summarize the CAUTION statement on page 14.3-4.

Example 2:
Study the solution for Example 2 part a on page 14.3-4 and record the answer below. Complete parts b and c on your own and check your answers by clicking on the link. If your answers are incorrect, watch the video to find your error.

Add.

a. $(y^2 + 3y + 7) + (y^2 - 3y - 2)$

b. $(10p^3 + 7p - 13) + (5p^2 - 4p)$

c. $(3m^3 + m^2 - 8) + (2m^3 - 4m^2 + 3m) + (5m^2 + 4)$

317

Topic 14.3

Topic 14.3 Objective 2: Find the Opposite of a Polynomial

In Topic 10.1 we learned that a _____ can be used to represent the _____ of a real number.

What is the "opposite" of the polynomial $x^2 - 5x + 7$?

What are **Opposite Polynomials**?

Summarize the TIP found on page 14.3-6.

Example 3:
Study the solutions for Example 3 parts a – c on page 14.3-6 and record the answers below.

Find the opposite of each polynomial

a. $x^2 + 6x + 8$ b. $8y - 27$ c. $-m^3 - 5m^2 + m + 7$

Topic 14.3 Objective 3: Subtract Polynomials

What is the procedure for **Subtracting Polynomials**?

Topic 14.3

Example 4:
Study the solution for Example 4 part a on page 14.3-8 and record the answer below. Complete part b on your own and check your answer by clicking on the link. If your answer is incorrect, watch the video to find your error.

Subtract.

a. $(9x + 13) - (6x - 4)$

b. $(3a^2 + 5a - 8) - (-2a^2 + a - 7)$

Topic 14.3

Topic 14.4

Topic 14.4 Guided Notebook
Topic 14.4 Multiplying Polynomials

Read the list of "THINGS TO KNOW" and review any concepts you are unfamiliar with.

Topic 14.4 Objective 1: Multiply Monomials

What is the procedure for **Multiplying Monomials**?

Example 1:
Study the solution for Example 1 part a on page 14.4-3 and record your answer below. Complete parts b and c on your own and check your answers by clicking on the popup.

Multiply.

a. $(6x^5)(7x^2)$ b. $\left(-\frac{3}{4}x^2\right)\left(-\frac{2}{9}x^8\right)$ c. $(3x^2)(-0.2x^3)$

Topic 14.4 Objective 2: Multiply a Polynomial by a Monomial

What is the procedure for **Multiplying Polynomials by Monomials**?

Example 2:
Study the solution for Example 2 part a on page 14.4-4 and record your answer below. Complete part b on your own and check your answer by clicking on the popup.

Multiply.

a. $3x(4x-5)$ b. $-4x^2(3x^2+x-7)$

Topic 14.4

Example 3:
Complete Example 3 parts a and b on page 14.4-5 on your own. Check your answers by clicking on the link. If your answers are incorrect, watch the video to find your error.

Multiply.

a. $\dfrac{1}{2}x^2(4x^2-6x+2)$

b. $0.25x^3(6x^3-10x^2+4x-7)$

Topic 14.4 Objective 3: Multiply Two Binomials

What is the procedure for **Multiplying Two Binomials**?

Example 4:
Study the solutions for Example 4 parts a and b on page 14.4-6 and record your answers below. Complete part c on your own and check your answer by clicking on the link. If your answer is incorrect, watch the video to find your error.

Multiply using the distributive property twice.

a. $(x+3)(x+2)$

b. $(x+6)(x-2)$

c. $(x-4)(x-5)$

Read and summarize the CAUTION statement on page 14.4-7.

Take notes on the animation on page 14.4-8 illustrating how to use the **FOIL method**.

Topic 14.4

Read and summarize the CAUTION statement on page 14.4-8.

Example 5:
Study the solution for Example 5 part a on page 14.4-8 and record your answer below. Complete parts b and c on your own and check your answers by clicking on the link. If your answers are incorrect, watch the video to find your error.

Multiply using the FOIL method.

a. $(x-4)(2x+3)$ b. $\left(\dfrac{1}{2}x-6\right)(3x-4)$ c. $(5x+7)(4x+3)$

Read and summarize the CAUTION statement on page 14.4-10.

Topic 14.4 Objective 4: Multiply Two or More Polynomials

Take notes on the animation on page 14.4-11 and explain the method for **Multiplying Two or More Polynomials**.

Example 6:
Study the solution for Example 6 part a on page 14.4-11 and record your answer below. Complete part b on your own and check your answer by clicking on the link. If your answer is incorrect, watch the video to find your error.

Multiply.

a. $(x+2)(2x^2-7x+3)$ b. $(y^2+2y-9)(2y^2-4y+7)$

Topic 14.4

Example 7:
Work through Example 7 on page 14.4-12 and record your answers below. Check your answers by clicking on the link. If your answers are incorrect, watch the video to find your error.

Multiply.

a. $-4x(2x-1)(x+3)$

b. $(x-1)(x+3)(3x-2)$

Topic 14.5 Guided Notebook

Topic 14.5 Special Products

Read the list of "THINGS TO KNOW" and review any concepts you are unfamiliar with.

Topic 14.5 Objective 1: Square a Binomial Sum

What is a **binomial sum**?

Copy the method for squaring the binomial sum, $(A+B)^2$.

What is the **Square of a Binomial Sum Rule**?

Read and summarize the CAUTION statement on page 14.5-3.

Example 1:
Study the solution for Example 1 parts a and b on page 14.5-4 and record the answers below. Complete parts c and d on your own and check your answers by clicking on the link. If your answers are incorrect, watch the video to find your error.

Multiply.

a. $(x+7)^2$
b. $(0.2m+1)^2$
c. $\left(z^2+\dfrac{1}{4}\right)^2$
d. $\left(10y+\dfrac{2}{5}\right)^2$

Topic 14.5

Topic 14.5 Objective 2: Square a Binomial Difference

What is a **binomial difference**?

Copy the method for squaring the binomial difference, $(A-B)^2$.

What is the **Square of a Binomial Difference Rule**?

Read and summarize the CAUTION statement on page 14.5-5.

Example 2:
Study the solutions for Example 2 parts a and b on page 14.5-6 and record the answers below. Complete parts c and d on your own and check your answers by clicking on the link. If your answers are incorrect, watch the video to find your error.

Multiply.

a. $(x-3)^2$ b. $\left(2z-\dfrac{1}{6}\right)^2$ c. $(w^3-0.7)^2$ d. $(5p-1.2)^2$

Topic 14.5

What are **perfect square trinomials**?

Topic 14.5 Objective 3: Multiply the Sum and Difference of Two Terms

What are **conjugates**?

What is the **Sum and Difference of Two Terms Rule (Product of Conjugates Rule)**?

Example 3:
Study the solution for Example 3 parts a and b on page 14.5-9 and record the answers below. Complete parts c and d on your own and check your answers by clicking on the link. If your answers are incorrect, watch the video to find your error.

Multiply.

a. $(x+4)(x-4)$ b. $\left(5y+\dfrac{1}{2}\right)\left(5y-\dfrac{1}{2}\right)$ c. $(8-x)(8+x)$ d. $(3z^2+0.5)(3z^2-0.5)$

Topic 14.5

Write down the three **Special Product Rules for Binomials**.

Topic 14.6

Topic 14.6 Guided Notebook
Topic 14.6 Negative Exponents and Scientific Notation

Read the list of "THINGS TO KNOW" and review any concepts you are unfamiliar with.

Topic 14.6 Objective 1: Use the Negative-Power Rule

What is the definition of a **Negative Exponent**?

Example 1:
Study the solutions for Example 1 parts a–c on page 14.6-4 and record the answers below. Complete parts d–f on your own and check your answers by clicking on the link. If your answers are incorrect, watch the video to find your error.

Write each expression with positive exponents. Then simplify if possible.

a. x^{-4} b. 2^{-3} c. $7x^{-3}$ d. $(-2)^{-4}$ e. -3^{-2} f. $2^{-1}+3^{-1}$

Read and summarize the CAUTION statement on page 14.6-5.

What is the **Negative-Power Rule**?

Example 2:
Study the solutions for Example 2 parts a–c on page 14.6-6 and record the answers below. Complete parts d–f on your own and check your answers by clicking on the link. If your answers are incorrect, watch the video to find your error.

Write each expression with positive exponents. Then simplify if possible.

a. $\dfrac{1}{y^{-5}}$ b. $\dfrac{1}{6^{-2}}$ c. $\dfrac{3}{4t^{-7}}$ d. $\dfrac{-8}{q^{-11}}$ e. $\dfrac{m^{-9}}{n^{-4}}$ f. $\dfrac{5^{-3}}{2^{-4}}$

Topic 14.6

Topic 14.6 Objective 2: Simplify Expressions Containing Negative Exponents Using a Combination of Rules

What are the four requirements for **simplified exponential expressions**?
-
-
-
-

Summarize the **Rules for Exponents**.

Product Rule

Quotient Rule

Zero-Power Rule

Power-to-Power Rule

Product-to-Power Rule

Quotient-to-Power Rule

Negative-Power Rule

Example 3:
Study the solutions for Example 3 parts a and b on page 14.6-9 and record the answers below. Complete parts c and d on your own and check your answers by clicking on the link. If your answers are incorrect, watch the video to find your error.

Simplify.

a. $(9x^{-5})(7x^2)$ b. $(p^{-4})^2$ c. $\dfrac{52m^{-4}}{13m^{-10}}$ d. $(w^{-1}z^3)^{-4}$

Topic 14.6

Example 4:
Study the solutions for Example 4 parts a and b on page 14.6-10 and record the answers below. Complete parts c and d on your own and check your answers by clicking on the link. If your answers are incorrect, watch the video to find your error.

Simplify.

a. $\dfrac{(3xz)^{-2}}{(2yz)^{-3}}$

b. $\left(\dfrac{10}{x}\right)^{-3}$

c. $\dfrac{\left(2a^5b^{-6}\right)^3}{4a^{-1}b^5}$

d. $\left(\dfrac{-5xy^{-3}}{x^{-2}y^5}\right)^4$

Topic 14.6 Objective 3: Convert a Number from Standard From to Scientific Notation

What is **Scientific Notation**?

What is the procedure for **Converting from Standard Form to Scientific Notation**?

1.

2.

Example 5:
Study the solutions for Example 5 parts a and b on page 14.6-13 and record the answers below. Complete parts c and d on your own and check your answers by clicking on the link. If your answers are incorrect, watch the video to find your error.

Write each number in scientific notation.

a. 56,800,000,000,000,000

b. 0.0000000467

c. 0.00009012

d. 200,000,000

Copyright © 2014 Pearson Education, Inc.

Topic 14.6

Topic 14.6 Objective 4: Convert a Number from Scientific Notation to Standard Form

What is the procedure for **Converting from Standard Form to Scientific Notation**?

1.

2.

Example 6:
Study the solutions for Example 6 parts a and b on page 14.6-15 and record the answers below. Complete parts c and d on your own and check your answers by clicking on the link. If your answers are incorrect, watch the video to find your error.

Write each number in standard form.

a. 4.98×10^{-5} b. 9.4×10^{7} c. -3.015×10^{9} d. 1.203×10^{-4}

Topic 14.6 Objective 5: Multiply and Divide with Scientific Notation

Example 7:
Study the solutions for Example 7 parts a and b on page 14.6-17 and record the answers below. Complete parts c and d on your own and check your answers by clicking on the link. If your answers are incorrect, watch the video to find your error.

Perform the indicated operations. Write your results in scientific notation..

a. $(1.8 \times 10^5)(3 \times 10^8)$ b. $\dfrac{2.16 \times 10^{12}}{4.5 \times 10^3}$

c. $(-7.4 \times 10^9)(6.5 \times 10^{-4})$ d. $\dfrac{5.7 \times 10^{-3}}{7.5 \times 10^{-7}}$

Topic 14.7 Guided Notebook

Topic 14.7 Dividing Polynomials

Read the list of "THINGS TO KNOW" and review any concepts you are unfamiliar with.

Topic 14.7 Objective 1: Divide Monomials

What is the procedure for **Dividing Monomials**?

Example 1:
Study the solution for Example 1 on page 14.7-3 and record the answers below.

Divide

a. $\dfrac{32x^7}{4x^3}$
b. $\dfrac{9y^4}{45y^4}$
c. $\dfrac{60y}{5y^4}$

Topic 14.7 Objective 2: Divide a Polynomial by a Monomial

What is the procedure for **Dividing Polynomials by Monomials**?

Example 2:
Study the solution for Example 2 part a on page 14.7-5 and record the answer below. Complete part b on your own and check your answer by clicking on the link. If your answer is incorrect, watch the video to find your error.

Divide.

a. $\dfrac{12x^3 - 28x^2}{4x^2}$
b. $(9m^5 - 15m^4 + 18m^3) \div 3m^3$

Topic 14.7

Example 3:
Study the solution for Example 3 on page 14.7-7 and record the answer below.

Divide $\dfrac{54t^3 - 12t^2 - 24t}{6t^2}$.

Topic 14.7 Objective 3: Divide Polynomials Using Long Division

Read and summarize the CAUTION statement on page 14.7-8.

What is the **Process for Polynomial Long Division**?

1.

2.

3.

4.

Example 4:
Study the solution for Example 4 on page 14.7-9 and record the answer below.

Divide $(2x^2 + x - 15) \div (x + 3)$

Topic 14.7

Example 5:
Complete Example 5 on page 14.7-12 on your own. Check your answer by clicking on the link. If your answer is incorrect, watch the video to find your error.

Divide $\dfrac{x^2 + 26x - 6x^3 - 12}{2x - 3}$

Example 6:
Complete Example 6 on page 14.7-14 on your own. Check your answer by clicking on the link. If your answer is incorrect, watch the video to find your error.

Divide $\dfrac{3t^3 - 11t - 12}{t + 4}$

Topic 14.7

Topic 14.8 Guided Notebook

Topic 14.8 Polynomials in Several Variables

Read the list of "THINGS TO KNOW" and review any concepts you are unfamiliar with.

Topic 14.8 Objective 1: Determine the Degree of a Polynomial in Several Variables

What is the definition of a **Polynomial in Several Variables**?

Example 1:
Study the solution for Example 1 part a on page 14.8-3 and record your answer below. Complete part b on your own and check your answer by viewing the popup,

Determine the coefficient and degree of each term; then find the degree of the polynomial.

a. $2x^3y - 7x^2y^3 + xy^2$
b. $3x^2yz^3 - 4xy^3z + xy^2z^4$

Topic 14.8 Objective 2: Evaluate Polynomials in Several Variables

Example 2:
Study the solution for Example 2 part a on page 14.8-5 and record your answer below. Complete part b on your own and check your answer by clicking on the link. If your answer is incorrect, watch the video to find your error.

a. Evaluate $3x^2y - 2xy^3 + 5$ for $x = -2$ and $y = 3$.

b. Evaluate $-a^3bc^2 + 5a^2b^2c - 2ab$ for $a = 2$, $b = -1$ and $c = 4$.

Topic 14.8

Topic 14.8 Objective 3: Add or Subtract Polynomials in Several Variables

We add polynomials in several variables by removing _____
and _____.

Example 3:
Study the solutions for Example 3 parts a and b on page 14.8-6 and record your answers below. Complete parts c and d on your own and check your answers by clicking on the link. If your answers are incorrect, watch the video to find your error.

Add or subtract as indicated.

a. $(2x^2 + 3xy - 7y^2) + (4x^2 - xy + 11y^2)$

b. $(4a^2 - 3ab + 2b^2) - (6a^2 - 5ab + 7b^2)$

c. $(7x^4 + 3x^3y^3 - 2xy^3 + 5) + (2x^4 - x^3y^3 + 8xy^3 - 10)$

d. $(10x^3y + 2x^2y^2 - 5xy^3 - 8) - (6x^3y + x^2y^2 - 3xy^3)$

Topic 14.8 Objective 3: Multiply Polynomials in Several Variables

We multiply polynomials in several variables _____
_____.

Example 4:
Study the solution for Example 4 on page 14.8-8 and record the answer below.

Multiply: $5xy^2\left(4x^2 - 3xy + 2y^2\right)$

Example 5:
Study the solution for Example 5 on page 14.8-9 and record the answer below.

Multiply: $(3x - 2y)(4x + 3y)$

Example 6:
Study the solution for Example 6 part a on page 14.8-10 and record your answer below. Complete parts b and c on your own and check your answers by clicking on the link. If your answers are incorrect, watch the video to find your answer.

Multiply:

a. $\left(6x^2 + 5y\right)^2$ b. $\left(4x^3 - 9y^2\right)^2$ c. $\left(2x^2y - 7\right)\left(2x^2y + 7\right)$

Topic 14.8

Example 7:
Study the solution for Example 7 on page 14.8-11 and record the answer below.

Multiply: $(x+2y)(x^2-4xy+y^2)$

Read and summarize the CAUTION statement on page 14.8-11.

Topic 15.1

Topic 15.1 Guided Notebook

Topic 15.1 Greatest Common Factor and Factoring by Grouping

Read the list of "THINGS TO KNOW" and review any concepts you are unfamiliar with.

Topic 15.1 Objective 1: Find the Greatest Common Factor of a Group of Integers

Write down the definitions for the following terms.
Factoring

Factored form

Factor (as a noun)
p

Factor (as a verb)

Factor over the integers

Greatest common factor

Record the steps for **Finding the GCF of a Group of Integers**.

1.

2.

3.

Example 1:
Study the solutions for Example 1 parts a and b on page 15.1-4 and record the answers below. Complete part c on your own and check your answer by clicking on the link. If your answer is incorrect, watch the video to find your error.

Find the GCF of each group of integers.

a. 36 and 60   b. 28 and 45   c. 75, 90, and 105

Topic 15.1

Topic 15.1 Objective 2: Find the Greatest Common Factor of a Group of Monomials

What is the **greatest common factor (GCF) of a group of monomials**?

What are the **Common Variable Factors for a GCF**?

Example 2:
Study the solutions for Example 2 parts a – c on page 15.1-6 and record the answers below.

Find the GCF of each group of exponential expressions.

a. x^4 and x^7 b. y^3, y^6, and y^9 c. w^6z^2, w^3z^5, and w^5z^4

What is the three-step process for **Finding the GCF of a Group of Monomials**?

1.

2.

3.

Example 3:
Study the solutions for Example 3 parts a and b on page 15.1-8 and record the answers below. Complete parts c and d on your own and check your answers by clicking on the link. If your answers are incorrect, watch the video to find your error.

Find the GCF of each group of monomials.

a. $14x^6$ and $21x^8$ b. $6a^2, 10ab$, and $14b^2$

c. $40x^5y^6, -48x^9y$, and $24x^2y^4$ d. $14m^3n^2, 6m^5n$, and $9m^4$

Topic 15.1 Objective 3: Factor Out the Greatest Common Factor from a Polynomial

What is the **greatest common factor (GCF) of a polynomial**?

Topic 15.1

What is the four-step process for **Factoring Out the GCF from a Polynomial**?

1.

2.

3.

4.

Example 4:
Study the solution for Example 4 part a on page 15.1-11 and record the answer below. Complete parts b and c on your own and check your answers by clicking on the link. If your answers are incorrect, watch the video to find your error.

Factor out the GCF from each binomial

a. $6x + 12$ b. $w^5 + w^4$ c. $8y^3 - 12y^2$

Read and summarize the CAUTION statement on page 15.1-13.

Example 5:
Study the solution for Example 5 part a on page 15.1-13 and record the answer below. Complete part b on your own and check your answer by clicking on the link. If your answer is incorrect, watch the video to find your error.

Factor out the GCF from each polynomial

a. $9p^5 + 18p^4 + 54p^3$ b. $10a^4b^6 - 15a^3b^7 + 35a^2b^8$

Example 6:
Study the solution for Example 6 on page 15.1-14.

Topic 15.1

Example 7:
Study the solution for Example 7 part a page 15.1-16 and record the answer below. Complete part b on your own and check your answer by clicking on the link. If your answer is incorrect, watch the video to find your error.

Factor out the common binomial factor as the GCF.

a. $4x(y+5) + 11(y+5)$

b. $7x(x+y) - (x+y)$

Topic 15.1 Objective 4: Factor by Grouping

What is the four-step process for **Factoring a Polynomial by Grouping**?

1.

2.

3.

4.

Read and summarize the CAUTION statement on page 15.1-18.

Example 8:
Study the solutions for Example 8 parts a and b on page 15.1-18. Complete parts c and d on your own and check your answers by clicking on the link. If your answers are incorrect, watch the video to find your error.

Factor by grouping.

c. $3m^2 + 3m - 2mn - 2n$

d. $4w^3 - 14w^2 - 10w + 35$

Read and summarize the CAUTION statement on page 15.1-20.

Topic 15.2 Guided Notebook

Topic 15.2 Factoring Trinomials of the Form $x^2 + bx + c$

Read the list of "THINGS TO KNOW" and review any concepts you are unfamiliar with.

Topic 15.2 Objective 1: Factor Trinomials of the Form $x^2 + bx + c$

Watch the animation about factoring trinomials on page 15.2-3 and take notes below.

Record the steps for **Factoring Trinomials of the Form** $x^2 + bx + c$.

1.

2.

3.

Example 1:
Study the solution for Example 1 part a on page 15.2-4 and record the answer below. Complete part b on your own and check your answer by clicking on the link. If your answer is incorrect, watch the video to find your error.

Factor each trinomial.

a. $x^2 + 11x + 18$ b. $x^2 + 13x + 30$

Topic 15.2

What is a **prime number**?

What is the definition of a **Prime Polynomial**?

Read and summarize the CAUTION statement on page 15.2-6.

Example 2:
Study the solution for Example 2 on page 15.2-7 and record your answer below.

Factor $x^2 + 14x + 20$

Example 3:
Study the solution for Example 3 part a on page 15.2-8 and record the answer below. Complete parts b and c on your own and check your answers by clicking on the link. If your answers are incorrect, watch the video to find your error.

Factor.

a. $x^2 - 13x + 40$

b. $m^2 - 5m - 36$

c. $w^2 + 7w - 60$

Topic 15.2 Objective 2: Factor Trinomials of the Form $x^2 + bxy + cy^2$

Example 4:
Study the solution for Example 4 part a on page 15.2-10 and record the answer below. Complete part b on your own and check your answer by clicking on the link. If your answer is incorrect, watch the video to find your error.

Factor.

a. $x^2 + 10xy + 24y^2$

b. $m^2 + 22mn - 48n^2$

Topic 15.2 Objective 3: Factor Trinomials of the Form $x^2 + bx + c$ after Factoring Out the GCF

When is a polynomial **factored completely**?

Example 5:
Study the solution for Example 5 part a on page 15.2-12 and record the answer below. Complete part b on your own and check your answer by clicking on the link. If your answer is incorrect, watch the video to find your error.

Factor completely.

a. $4x^2 - 28x - 32$

b. $2y^3 - 36y^2 + 64y$

Topic 15.2

Example 6:
Complete Example 6 on page 15.1-14 on your own. Check your answer by clicking on the link. If your answer is incorrect, watch the video to find your error.

Factor $-x^2 + 3x + 10$

Topic 15.3 Guided Notebook

Topic 15.3 Factoring Trinomials of the Form $ax^2 + bx + c$ Using Trial and Error

Read the list of "THINGS TO KNOW" and review any concepts you are unfamiliar with.

Topic 15.3 Objective 1: Factor Trinomials of the Form $ax^2 + bx + c$ Using Trial and Error

Watch the animation about factoring trinomials on page 15.3-3 and take notes below.

What is the four-step strategy for **Factoring Trinomials of the Form $ax^2 + bx + c$?**

  1.

  2.

  3.

  4.

Example 1:
Study the solution for Example 1 on page 15.3-5 and record the answer below.

Factor $3x^2 + 7x + 2$.

Read and summarize the CAUTION on page 15.3-6.

Topic 15.3

Example 2:
Study the solution for Example 2 on page 15.3-6 and record the answer below.

Factor $5x^2 + 17x + 6$.

Example 3:
Complete Example 3 parts a and b on page 15.3-8 on your own. Check your answers by clicking on the link. If your answers are incorrect, watch the video to find your error.

Factor.

a. $4x^2 - 5x - 6$

b. $12n^2 - 16n + 5$

Read and summarize the CAUTION statement on page 15.3-9.

Example 4:
Study the solution for Example 4 on page 15.3-10 and record the answer below.

Factor $2y^2 - 19y + 15$.

Topic 15.3 Objective 2: Factor Trinomials of the Form $ax^2 + bxy + cy^2$ Using Trial and Error

Example 5:
Study the solution for Example 5 part a on page 15.3-12 and record the answer below. Complete part b on your own and check your answer by clicking on the link. If your answer is incorrect, watch the video to find your error.

Factor.

a. $6x^2 + 17xy - 3y^2$

b. $2m^2 + 11mn + 12n^2$

Topic 15.4

Topic 15.4 Guided Notebook

Topic 15.4 Factoring Trinomials of the Form $ax^2 + bx + c$ Using the ac Method

Read the list of "THINGS TO KNOW" and review any concepts you are unfamiliar with.

Topic 15.4 Objective 1: Factor Trinomials of the Form $ax^2 + bx + c$ Using the ac Method

Record the steps for **The ac Method for Factoring Trinomials of the Form $ax^2 + bx + c$**

1.

2.

3.

4.

5.

Watch the animation about factoring trinomials on page 15.4-3 and take notes below.

What are the two other names for the ac Method?

Topic 15.4

Example 1:
Study the solution for Example 1 on page 15.4-4 and record the answer below..

Factor $3x^2 + 14x + 8$ using the *ac* method.

Example 2:
Study the solution for Example 2 on page 15.4-6 and record the answer below.

Factor $2x^2 - 3x - 20$ using the *ac* method.

Example 3:
Complete Example 3 parts a – c on page 15.4-7 on your own. Check your answers by clicking on the link. If your answers are incorrect, watch the video to find your error.

Factor each trinomial using the *ac* method. If the trinomial is prime, state this as your answer.

a. $2x^2 + 9x - 18$

b. $6x^2 - 23x + 20$

c. $5x^2 + x + 6$

Topic 15.4 Objective 2: Factor Trinomials of the Form $ax^2 + bxy + cy^2$ Using the *ac* Method

Example 4:
Complete Example 4 on page 15.4-8 on your own. Check your answer by clicking on the link. If your answer is incorrect, watch the video to find your error.

Factor $2p^2 + 7pq - 15q^2$ using the *ac* method.

Topic 15.4 Objective 3: Factor Trinomials of the Form $ax^2 + bx + c$ after Factoring out the GCF

Example 5:
Complete Example 5 on page 15.4-10 on your own. Check your answer by clicking on the link. If your answer is incorrect, watch the video to find your error.

Factor completely: $24t^5 - 52t^4 - 20t^3$

Example 6:
Complete Example 6 on page 15.4-11 on your own. Check your answer by clicking on the link. If your answer is incorrect, watch the video to find your error.

Factor completely: $-2x^2 + 9x + 35$

Topic 15.4

Topic 15.5

Topic 15.5 Guided Notebook
Topic 15.5 Factoring Special Forms

Read the list of "THINGS TO KNOW" and review any concepts you are unfamiliar with.

Topic 15.5 Objective 1: Factor the Difference of Two Squares

Summarize **Factoring the Difference of Two Squares**.

Read and summarize the CAUTION statement on page 15.5-4.

Example 1:
Study the solutions for Example 1 parts a and b on page 15.5-4 and record the answers below.

Factor each expression completely.

a. $x^2 - 9$
b. $16 - y^2$

What is a **perfect square**? Give two examples of a perfect square.

Example 2:
Study the solution for Example 2 part a on page 15.5-6 and record the answer below. Complete parts b–d on your own and check your answers by clicking on the link. If your answers are incorrect, watch the video to find your error.

Factor each expression completely.

a. $z^2 - \dfrac{25}{16}$
b. $36x^2 - 25$
c. $4 - 49n^6$
d. $81m^2 - n^2$

Topic 15.5

Example 3:
Study the solution for Example 3 part a on page 15.5-7 and record the answer below. Complete part b on your own and check your answer by clicking on the link. If your answer is incorrect, watch the video to find your error.

Factor each expression completely.

a. $3x^2 - 75$

b. $36x^3 - 64x$

Read and summarize the CAUTION statement on page 15.5-8.

Example 4:
Study the solution for Example 4 on page 15.5-9 and record the answer below.

Factor completely.

$16x^4 - 81$

Topic 15.5 Objective 2: Factor Perfect Square Trinomials

Summarize **Factoring Perfect Square Trinomials.**

Example 5:
Study the solutions for Example 5 parts a and b on page 15.5-11 and record the answers below.

Factor each expression completely.

a. $x^2 + 6x + 9$

b. $y^2 - 10y + 25$

Topic 15.5

Example 6:
Study the solution for Example 6 part a on page 15.5-12 and record the answer below. Complete part b on your own and check your answer by clicking on the link. If your answer is incorrect, watch the video to find your error.

Factor each expression completely.

a. $4x^2 + 12x + 9$

b. $25y^2 - 60y + 36$

Example 7:
Complete Example 7 parts a and b on page 15.5-13 on your own. Check your answers by clicking on the link. If your answers are incorrect, watch the video to find your error.

Factor each expression completely.

a. $16x^2 + 24xy + 9y^2$

b. $m^4 - 12m^2 + 36$

Topic 15.5 Objective 3: Factor the Sum or Difference of Two Cubes

Summarize **Factoring the Sum and Difference of Two Cubes**

What are **perfect cubes**? Give two examples of a perfect cube.

Topic 15.5

Example 8:
Study the solutions for Example 8 parts a and b on page 15.5-16 and record the answers below.

Factor each expression completely.

a. $x^3 + 64$

b. $z^3 - 8$

Example 9:
Study the solution for Example 9 part a on page 15.5-18 and record the answer below. Complete parts b and c on your own and check your answers by clicking on the link. If your answers are incorrect, watch the video to find your error.

Factor each expression completely.

a, $125y^3 - 1$

b. $128z^3 + 54y^3$

c. $8x^3y^3 + y^5$

Topic 15.6 Guided Notebook

Topic 15.6 A General Factoring Strategy

Read the list of "THINGS TO KNOW" and review any concepts you are unfamiliar with.

Topic 15.6 Objective 1: Factor Polynomials Completely

What is the four-step **General Strategy for Factoring Polynomials Completely**?

1.

2.
   a.

   b.

   c.

3.

4.

Topic 15.6

Example 1:
Study the solutions for Example 1 parts a and b on page 15.6-4 and record the answers below. Complete parts c and d on your own and check your answers by clicking on the link. If your answers are incorrect, watch the video to find your error.

Factor each expression completely.

a. $w^2 - w - 20$

b. $4y^4 - 32y$

c. $x^2 - 14x + 49$

d. $3z^3 - 15z^2 - 42z$

Example 2:
Study the solution for Example 2 part a on page 15.6-7 and record the answer below. Complete parts b and c on your own and check your answers by clicking on the link. If your answers are incorrect, watch the video to find your error.

Factor each expression completely.

a. $2x^3 - 5x^2 - 8x + 20$

b. $3a^2 - 10a - 8$

c. $3z^2 + z - 1$

Copyright © 2014 Pearson Education, Inc.

Topic 15.6

Example 3:
Work through Example 3 part a on page 15.6-8 and record the answer below. Complete parts b–d on your own and check your answers by clicking on the link. If your answers are incorrect, watch the video to find your error.

Factor each expression completely.

a. $10x^2 + 11xy - 6y^2$

b. $2p^2 - 32pq + 128q^2$

c. $7x^2z - 14x$

d. $-3y^4z - 24yz^4$

Topic 15.6

Topic 15.7 Guided Notebook

Topic 15.7 Solving Polynomial Equations by Factoring

Read the list of "THINGS TO KNOW" and review any concepts you are unfamiliar with.

Topic 15.7 Objective 1: Solve Quadratic Equations by Factoring

Write down the definitions for the following terms.

Polynomial equation

Standard form

Degree of a polynomial equation

Write the definition of a **Quadratic Equation**.

What is the **Zero Product Property**?

Read and summarize the CAUTION statement on page 15.7-4.

Topic 15.7

Example 1:
Study the solution for Example 1 part a on page 15.7-5 and record the answer below. Complete part b on your own and check your answer by clicking on the link. If your answer is incorrect, watch the video to find your error.

Solve each equation.

a. $(x + 10)(x - 3) = 0$

b. $x(3x + 5) = 0$

What is the four-step process for **Solving Polynomial Equations by Factoring**?

1.

2.

3.

4.

Example 2:
Study the solution for Example 2 part a on page 15.7-6 and record the answer below. Complete part b on your own and check your answer by clicking on the link. If your answer is incorrect, watch the video to find your error.

Solve each equation by factoring.

a. $z^2 + 4z - 12 = 0$

b. $-4x^2 + 28x - 40 = 0$

Topic 15.7

Read and summarize the CAUTION statement on page 15.7-8.

Example 3:
Study the solution for Example 3 part a on page 15.7-8 and record the answer below. Complete parts b and c on your own and check your answers by clicking on the link. If your answers are incorrect, watch the video to find your error.

Solve each equation by factoring.

a. $9w^2 + 64 = 48w$ b. $4m^2 = 49$ c. $3x(x - 2) = 2 - x$

Read and summarize the CAUTION statement on page 15.7-10.

Example 4:
Complete Example 4 on page 15.7-11 parts a and b on your own. Check your answers by clicking on the link. If your answers are incorrect, watch the video to find your error.

Solve each equation by factoring.

a. $(x + 2)(x - 5) = 18$ b. $(x + 3)(3x - 5) = 5(x + 1) - 10$

Read and summarize the CAUTION statement on page 15.7-12.

Topic 15.7

Topic 15.7 Objective 2: Solve Polynomial Equations by Factoring

Example 5:
Study the solution for Example 5 part a on page 15.7-13 and record the answer below. Complete parts b and c on your own and check your answers by clicking on the link. If your answers are incorrect, watch the video to find your error.

Solve each equation by factoring.

a. $(x + 7)(2x - 1)(5x + 4) = 0$

b. $24x^3 + 8x^2 = 100x^2 - 28x$

c. $z^3 + z^2 = z + 1$

Example 6:
Study the solution for Example 6 on page 15.7-15 and record the answer below.

Solve by factoring.

$(2x - 9)(3x^2 - 16x - 12) = 0$

Topic 15.8 Guided Notebook

Topic 15.8 Applications of Quadratic Equations

Read the list of "THINGS TO KNOW" and review any concepts you are unfamiliar with.

Topic 15.8 Objective 1: Solve Application Problems Involving Consecutive Numbers

Review the **problem-solving strategy** on page 15.8-3.

Read and summarize the CAUTION statement on page 15.8-3.

Example 1:
Study the solution for Example 1 on page 15.8-3 and record the answer below.

The house numbers on the west side of a street are consecutive positive odd integers. The product of the house numbers for two next-door-neighbors on the west side of the street is 575. Find the house numbers.

Topic 15.8 Objective 2: Solve Application Problems Involving Geometric Figures

Example 2:
Complete Example 2 on page 15.8-5 on your own. Check your answer by clicking on the link. If your answer is incorrect, watch the video to find your error.

A swimming pool is 20 feet wide and 30 feet long. A sidewalk border around the pool has uniform width and an area that is equal to the area of the pool. Find the width of the border.

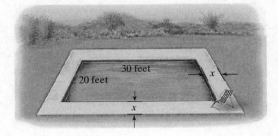

Topic 15.8

Topic 15.8 Objective 3: Solve Application Problems Using the Pythagorean Theorem

Write down the definition for the following terms.

Right triangles

Right angle

Hypotenuse

Legs

What is the **Pythagorean Theorem**?

Example 3:
Complete Example 3 on page 15.8-8 on your own. Check your answer by clicking on the link. If your answer is incorrect, watch the video to find your error.

A wire is attached to a cell phone tower for support. The length of the wire is 40 meters less than twice the height of the tower. The wire is fixed to the ground at a distance that is 40 meters more than the height of the tower. Find the length of the wire.

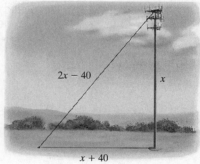

Topic 15.8

Read and summarize the CAUTION statement on page 15.8-9.

Topic 15.8 Objective 4: Solve Application Problems Involving Quadratic Models

Example 4:
Study the solution for Example 4 on page 15.8-10 and record the answer below. Watch the video for a detailed solution.

The Grand Canyon Skywalk sits 4000 ft above the Colorado River. If an object is dropped from the observation deck, its height h, in feet after t seconds, is given by

$$h = -16t^2 + 4000.$$

How long will it take for the object to be 400 feet above the Colorado River?

Example 5:
Complete Example 5 on page 15.8-12 on your own. Check your answer by clicking on the link. If your answer is incorrect, watch the video to find your error.

For household incomes under $100,000, the relationship between the percentage of households with home broadband access and the annual household income can be approximated by the model,

$$y = -0.01x^2 + 1.7x + 9.5.$$

Here, x is the annual household income (in $1000s) and y is the percentage of households with home broadband access. Use the model to estimate the annual household income if 75.5 percent of such households have home broadband access.

Topic 16.1 Guided Notebook

Topic 16.1 Simplifying Rational Expressions

Read the list of "THINGS TO KNOW" and review any concepts you are unfamiliar with.

Topic 16.1 Objective 1: Evaluate Rational Expressions

Write down the definition for a **Rational Expression** and give two examples.

Example 1:
Study the solutions for Example 1 parts a and b on page 16.1-4 and record the answers below.

Evaluate $\dfrac{x+8}{x-2}$ for the given value of x.

a. $x = 4$
b. $x = -6$

Example 2:
Study the solution for Example 2 on page 16.1-5 and record the answer below.

Evaluate $\dfrac{x^2 - y}{9x + 5y}$ for $x = 3$ and $y = -1$.

Topic 16.1 Objective 2: Find Restricted Values for Rational Expressions

Write down the definition of a **Restricted Value**.

Topic 16.1

What is the technique for **Finding Restricted Values for Rational Expressions in One Variable**?

Example 3:
Study the solution for Example 3 part a on page 16.1-7 and record the answer below. Complete part b on your own and check your answer by clicking on the link. If your answer is incorrect, watch the video to find your error.

Find any restricted values for each rational expression.

a. $\dfrac{3x+5}{3x-2}$

b. $\dfrac{x^2+2x-35}{x^2+x-30}$

Example 4:
Study the solutions for Example 4 parts a and b on page 16.1-8.

Topic 16.1 Objective 3: Simplify Rational Expressions

When is a fraction written in **lowest terms** or **simplest form**?

What is the **Simplification Principle for Rational Expressions**?

What are the three steps for **Simplifying Rational Expressions**?

1.

2

3.

Topic 16.1

Example 5:
Study the solution for Example 5 on page 16.1-11 and record the answer below.

Simplify $\dfrac{2x^2 - 6x}{7x - 21}$

Example 6:
Study the solution for Example 6 on page 16.1-12 and record the answer below.

Simplify $\dfrac{5x}{x^2 + 5x}$

Read and summarize the CAUTION statements on pages 16.1-12 and 16.1-13.

Example 7:
Complete Example 7 on page 16.1-13 on your own. Check your answer by clicking on the link. If your answer is incorrect, watch the video to find your error.

Simplify $\dfrac{y^2 + 2y - 24}{y^2 + 4y - 32}$

Example 8:
Complete Example 8 on page 16.1-14 on your own. Check your answer by clicking on the link. If your answer is incorrect, watch the video to find your error.

Simplify $\dfrac{2m^2 + m - 15}{2m^3 - 5m^2 - 18m + 45}$

Topic 16.1

Example 9:
Complete Example 9 on page 16.1-14 on your own. Check your answer by clicking on the link. If your answer is incorrect, watch the video to find your error.

Simplify $\dfrac{x^2 - xy - 12y^2}{2x^2 + 7xy + 3y^2}$

Example 10:
Study the solution for Example 10 on page 16.1-15 and record the answer below.

Simplify $\dfrac{w^2 - y^2}{2xy + 2xw}$

Example 11:
Study the solution for Example 11 on page 16.1-16 and record the answer below.

Simplify $\dfrac{3x - 10}{10 - 3x}$

Example 12:
Complete Example 12 on page 16.1-17 on your own. Check your answer by clicking on the link. If your answer is incorrect, watch the video to find your error.

Simplify $\dfrac{2x^2 - 27x + 70}{49 - 4x^2}$

Topic 16.2

Topic 16.2 Guided Notebook

Topic 16.2 Multiplying and Dividing Rational Expressions

Read the list of "THINGS TO KNOW" and review any concepts you are unfamiliar with.

Topic 16.2 Objective 1: Multiply Rational Expressions

What are the three steps for **Multiplying Rational Expressions**?

1.

2.

3.

Read and summarize the CAUTION statement on page 16.2-4.

Example 1:
Study the solution for Example 1 on page 16.2-5 and record the answer below.

Multiply $\dfrac{5x^2}{2y} \cdot \dfrac{6y^2}{25x^3}$

Example 2:
Study the solution for Example 2 on page 16.2-6 and record the answer below.

Multiply $\dfrac{3x-6}{2x} \cdot \dfrac{8}{5x-10}$

Example 3:
Complete Example 3 on page 16.2-7 on your own. Check your answer by clicking on the link. If your answer is incorrect, watch the video to find your error.

Multiply $\dfrac{x^2-4}{x^2+2x-35} \cdot \dfrac{x^2-25}{x+2}$

377
Copyright © 2014 Pearson Education, Inc.

Topic 16.2

Example 4:
Complete Example 4 on page 16.2-8 on your own. Check your answer by clicking on the link. If your answer is incorrect, watch the video to find your error.

Multiply $\dfrac{2x^2+3x-2}{3x^2-2x-1} \cdot \dfrac{3x^2+4x+1}{2x^2+x-1}$

Example 5:
Study the solution for Example 5 on page 16.2-8 and record the answer below.

Multiply $\dfrac{3x^2+9x+27}{x-1} \cdot \dfrac{x+3}{x^3-27}$

Example 6:
Complete Example 6 on page 16.2-10 on your own. Check your answer by clicking on the link. If your answer is incorrect, watch the video to find your error.

Multiply $\dfrac{3x^2+10x-8}{2x-3x^2} \cdot \dfrac{4x+1}{x+4}$

Example 7:
Complete Example 7 on page 16.2-11 on your own. Check your answer by clicking on the link. If your answer is incorrect, watch the video to find your error.

Multiply $\dfrac{x^2+xy}{3x+y} \cdot \dfrac{3x^2+7xy+2y^2}{x^2-y^2}$

Topic 16.2 Objective 2: Divide Rational Expressions

What is the two-step process for **Dividing Rational Expressions**?

    1.

    2.

Example 8:
Study the solutions for Example 8 parts a and b on page 16.2-13 and record the answers below.

Divide each rational expression.

a. $\dfrac{6x^5}{9y^3} \div \dfrac{5x^4}{3y^2}$

b. $\dfrac{(x+2)(x-1)}{(3x-5)} \div \dfrac{(x-1)(x+4)}{(2x+3)}$

Example 9:
Complete Example 9 on page 16.2-15 on your own. Check your answer by clicking on the link. If your answer is incorrect, watch the video to find your error.

Divide $\dfrac{9y^2 - 81}{4y^2} \div \dfrac{y+3}{8}$

Topic 16.2

Example 11:

Complete Example 11 on page 16.2-16 on your own. Check your answer by clicking on the link. If your answer is incorrect, watch the video to find your error.

Divide $\dfrac{x^3-8}{2x^2-x-6} \div \dfrac{x^2+2x+4}{6x^2+11x+3}$

Example 13:

Complete Example 13 on page 16.2-18 on your own. Check your answer by clicking on the link. If your answer is incorrect, watch the video to find your error.

Perform the indicated operations.

$\dfrac{x^2+2x-15}{x^2+2x-8} \cdot \dfrac{x^2+3x+2}{x^2+4x-21} \div \dfrac{x+2}{x^2+9x+14}$

Topic 16.3 Guided Notebook

Topic 16.3 Least Common Denominators

Read the list of "THINGS TO KNOW" and review any concepts you are unfamiliar with.

Topic 16.3 Objective 1: Find the Least Common Denominator of Rational Expressions

View the popup on page 16.3-3 and take notes on the steps for finding the LCD of rational numbers.

What are the three steps for **Finding the Least Common Denominator (LCD) of Rational Expressions**?

    1.

    2.

    3.

Read and summarize the CAUTION statement on page 16.3-4.

Example 1:
Study the solutions for Example 1 parts a and b on page 16.3-4 and record the answers below.

Find the LCD of the rational expressions.

  a. $\dfrac{7}{10x^3}, \dfrac{3}{5x^2}$ b. $\dfrac{x+2}{3x}, \dfrac{x-1}{2x^2+6x}$

Topic 16.3

Example 2:
Complete Example 2 on page 16.3-6 parts a and b on your own. Check your answers by clicking on the link. If your answers are incorrect, watch the video to find your error.

Find the LCD of the rational expressions.

a. $\dfrac{z^2}{6-z}, \dfrac{9}{2z-12}$

b. $\dfrac{y+2}{y^2+2y-3}, \dfrac{2y}{y^2+5y+6}$

Example 3:
Complete Example 3 on page 16.3-7 parts a and b on your own. Check your answers by clicking on the link. If your answers are incorrect, watch the video to find your error.

Find the LCD of the rational expressions.

a. $\dfrac{4x}{10x^2-7x-12}, \dfrac{2x-3}{5x^2-11x-12}$

b. $\dfrac{10-x}{6x^2+5x+1}, \dfrac{-4}{9x^2+6x+1}, \dfrac{x^2-7x}{10x^2-x-3}$

Topic 16.3 Objective 2: <u>Write Equivalent Rational Expressions</u>

When writing equivalent fractions, what is the key question that should be asked?

Topic 16.3

What is the three-step process for **Writing Equivalent Rational Expressions**?

1.

2.

3.

Read and summarize the CAUTION statement on page 16.3-9.

Example 4:
Study the solutions for Example 4 parts a and b on page 16.3-10 and record the answers below.

Write each rational expression as an equivalent rational expression with the desired denominator.

a. $\dfrac{3}{2x} = \dfrac{}{10x^3}$

b. $\dfrac{x+2}{3x+15} = \dfrac{}{3(x-1)(x+5)}$

Example 5:
Complete Example 5 on page 16.3-11 parts a and b on your own. Check your answers by clicking on the link. If your answers are incorrect, watch the video to find your error.

Write each rational expression as an equivalent rational expression with the desired denominator.

a. $\dfrac{-7}{1-4y} = \dfrac{}{8y^2-2y}$

b. $\dfrac{5z}{z^2+z-6} = \dfrac{}{(z-4)(z-2)(z+3)}$

Topic 16.3

Topic 16.4

Topic 16.4 Guided Notebook

Topic 16.4 Adding and Subtracting Rational Expressions

Read the list of "THINGS TO KNOW" and review any concepts you are unfamiliar with.

Topic 16.4 Objective 1: Add and Subtract Rational Expressions with Common Denominators

Record the technique for **Adding and Subtracting Rational Expressions with Common Denominators**.

Example 1:
Study the solutions for Example 1 parts a and b on page 16.4-4 and record the answers below.

Add or subtract.

a. $\dfrac{4z}{3} + \dfrac{5z}{3}$

b. $\dfrac{3r}{7s^2} - \dfrac{2r}{7s^2}$

Read and summarize the CAUTION statement on page 16.4-5.

Example 2:
Complete Example 2 on page 16.4-6 parts a and b on your own. Check your answers by clicking on the link. If your answers are incorrect, watch the video to find your error.

Add or subtract.

a. $\dfrac{9x}{x-4} + \dfrac{7x-2}{x-4}$

b. $\dfrac{5y+1}{y-2} - \dfrac{2y+3}{y-2}$

Topic 16.4

Example 3:
Complete Example 3 on page 16.4-8 parts a – c on your own. Check your answers by clicking on the link. If your answers are incorrect, watch the video to find your error.

Add or subtract.

a. $\dfrac{4}{x^2+2x-8}+\dfrac{x}{x^2+2x-8}$

b. $\dfrac{x}{x+2}-\dfrac{x-3}{x+2}$

c. $\dfrac{x^2-2}{x-5}-\dfrac{4x+3}{x-5}$

Topic 16.4 Objective 2: Add and Subtract Rational Expressions with Unlike Denominators

Record the steps for **Adding and Subtracting Rational Expressions with Unlike Denominators**?

1.

2.

3.

4.

Example 4:
Study the solution for Example 4 part a on 16.4-10 and record the answer below. Complete part b on your own and check your answer by clicking on the link. If your answer is incorrect, watch the video to find your error.

Perform the indicated operations and simplify.

a. $\dfrac{7}{6x}+\dfrac{3}{2x^3}$

b. $\dfrac{3x}{x-3}-\dfrac{x-2}{x+3}$

Topic 16.4

Example 5:
Study the solution for Example 5 part a on 16.4-13 and record the answer below. Complete part b on your own and check your answer by clicking on the link. If your answer is incorrect, watch the video to find your error.

Perform the indicated operations and simplify.

a. $\dfrac{z+2}{3z} - \dfrac{5}{3z+12}$

b. $\dfrac{5}{4m-12} + \dfrac{3}{2m}$

Example 6:
Study the solution for Example 6 part a on 16.4-15 and record the answer below. Complete part b on your own and check your answer by clicking on the link. If your answer is incorrect, watch the video to find your error.

Perform the indicated operations and simplify.

a. $2 + \dfrac{4}{x-5}$

b. $\dfrac{x^2-2}{x^2+6x+8} - \dfrac{x-3}{x+4}$

Topic 16.4

Example 7:
Complete Example 7 on page 16.4-18 parts a and b on your own. Check your answers by clicking on the link. If your answers are incorrect, watch the video to find your error.

Perform the indicated operations and simplify.

a. $\dfrac{x+7}{x^2-9} + \dfrac{3}{x+3}$

b. $\dfrac{x+1}{2x^2+5x-3} - \dfrac{x}{2x^2+3x-2}$

Example 8:
Study the solution for Example 8 on page 16.4-19 and record the answer below.

Perform the indicated operations and simplify.

$\dfrac{2y}{y-5} + \dfrac{y-1}{5-y}$

Example 9:
Complete Example 9 on page 16.4-20 on your own. Check your answer by clicking on the link. If your answer is incorrect, watch the video to find your error.

Perform the indicated operations and simplify.

$\dfrac{x+1}{x^2-6x+9} + \dfrac{3}{x-3} - \dfrac{6}{x^2-9}$

Topic 16.5

Topic 16.5 Guided Notebook

Topic 16.5 Complex Rational Expressions

Read the list of "THINGS TO KNOW" and review any concepts you are unfamiliar with.

Topic 16.5 Objective 1: <u>Simplify Complex Rational Expressions by First Simplifying the Numerator and Denominator</u>

Write down the definition of a **Complex Rational Expression**. Provide one example.

What form is a **simplified complex rational expression** written in?

Method I results from recognizing that the _____ is a division symbol. So, we _____ the numerator by the denominator.

Example 1:
Study the solution for Example 1 on page 16.5-4 and record the answer below.

Simplify $\dfrac{\dfrac{2}{9x}}{\dfrac{5}{6xy}}$

What are the three steps for **Method I for Simplifying Complex Rational Expressions**?

    1.

    2.

    3.

Topic 16.5

Example 2:
Study the solution for Example 2 part a on page 16.5-6 and record the answer below. Complete part b on your own and check your answer by clicking on the link. If your answer is incorrect, watch the video to find your error.

Use Method I to simplify each complex rational expression.

a. $\dfrac{\dfrac{1}{3}-\dfrac{1}{x}}{\dfrac{1}{9}-\dfrac{1}{x^2}}$

b. $\dfrac{4-\dfrac{5}{x-1}}{\dfrac{6}{x-1}-7}$

Topic 16.5 Objective 2: Simplify Complex Rational Expressions by Multiplying by a Common Denominator

Example 3:
Study the solution for Example 3 on page 16.5-9 and record the answer below.

Simplify $\dfrac{\dfrac{2}{9x}}{\dfrac{5}{6xy}}$

What is the three-step process for **Method II for Simplifying Complex Rational Expressions**

1.

2.

3.

Example 4:
Study the solution for Example 4 part a on page 16.5-11 and record the answer below. Complete part b on your own and check your answer by clicking on the link. If your answer is incorrect, watch the video to find your error.

Use Method II to simplify each complex rational expression.

a. $\dfrac{\dfrac{1}{3}-\dfrac{1}{x}}{\dfrac{1}{9}-\dfrac{1}{x^2}}$

b. $\dfrac{4-\dfrac{5}{x-1}}{\dfrac{6}{x-1}-7}$

Which of the two methods do you prefer? Why?

Topic 16.5

Example 5:

Complete Example 5 on page 16.5-14 on your own. Check your answer by clicking on the link. If your answer is incorrect, watch the video to find your error.

Simplify the complex rational expressions using Method I or Method II.

$$\dfrac{\dfrac{5}{n-2}-\dfrac{3}{n}}{\dfrac{6}{n^2-2n}+\dfrac{2}{n}}$$

Example 6:

Complete Example 6 on page 16.5-15 on your own. Check your answer by clicking on the link. If your answer is incorrect, watch the video to find your error.

Simplify the complex rational expression.

$$\dfrac{1-9y^{-1}+14y^{-2}}{1+3y^{-1}-10y^{-2}}$$

Topic 16.6

Topic 16.6 Guided Notebook
Topic 16.6 Solving Rational Equations

Read the list of "THINGS TO KNOW" and review any concepts you are unfamiliar with.

Topic 16.6 Objective 1: Identify Rational Equations

What is the definition of a **Rational Equation**?

Example 1:
Study the solutions for Example 1 parts a – d on page 16.6-3 and record the answers below.

Determine if each statement is a rational equation. If not, state why.

a. $\dfrac{x-4}{x} + \dfrac{4}{x+5} = \dfrac{6}{x}$

b. $\dfrac{5}{y} + \dfrac{7}{y+2}$

c. $\dfrac{\sqrt{k+1}}{k+3} = \dfrac{k-5}{k+4}$

d. $5n^{-1} = 3n^{-2}$

Topic 16.6 Objective 2: Solve Rational Equations

Example 2:
Study the solution for Example 2 part a on page 16.6-5 and record the answer below. Complete part b on your own and check your answer by clicking on the link. If your answer is incorrect, watch the video to find your error.

Solve.

a. $\dfrac{1}{2}x + \dfrac{2}{3} = \dfrac{3}{4}$

b. $\dfrac{1}{x} + \dfrac{1}{2} = \dfrac{1}{3}$

Topic 16.6

What are extraneous solutions?

What is the five-step process for Solving Rational Equations?
1.

2.

3.

4.

5.

Example 3:
Study the solution for Example 3 on page 16.6-8 and record the answer below.

Solve $\dfrac{2}{x} - \dfrac{x-3}{2x} = 3$

Example 4:
Complete Example 4 on page 16.6-10 on your own. Check your answer by clicking on the link. If your answer is incorrect, watch the video to find your error.

Solve $\dfrac{4}{5} - \dfrac{3}{x-3} = \dfrac{1}{x}$

Example 5:
Complete Example 5 on page 16.6-11 on your own. Check your answer by clicking on the link. If your answer is incorrect, watch the video to find your error.

Solve $\dfrac{m}{m+2} + \dfrac{5}{m-2} = \dfrac{20}{m^2-4}$

Example 6:
Work through Example 6 on page 16.6-13 and record the answer below. Check your answer by clicking on the link. If your answer is incorrect, watch the video to find your error.

Solve $\dfrac{2}{x-3} - \dfrac{4}{x^2-2x-3} = \dfrac{1}{x+1}$

Topic 16.6 Objective 3: Identify and Solve Proportions

Write down the definition for the following terms.

Ratio

Proportion

Cross-multiplying

Topic 16.6

Example 7:
Study the solution for Example 7 part a on page 16.6-15 and record the answer below. Complete part b on your own and check your answer by clicking on the link. If your answer is incorrect, watch the video to find your error.

Solve.

a. $\dfrac{8}{x+3} = \dfrac{5}{x}$
b. $\dfrac{x}{6} = \dfrac{2}{x-1}$

Read and summarize the CAUTION statement on page 16.6-16.

Topic 16.6 Objective 4: Solve a Formula Containing Rational Expressions for a Given Variable

What is a **formula**?

Example 8:
Study the solution for Example 8 part a on page 16.6-17 and record the answer below. Complete part b on your own. Watch the video to see the complete solution.

Solve each formula for the given variable.

a. $I = \dfrac{E}{r+R}$ for R
b. $\dfrac{1}{f} = \dfrac{1}{c} + \dfrac{1}{d}$ for d

Read and summarize the CAUTION statement on page 16.6-18.

Topic 16.7 Guided Notebook

Topic 16.7 Applications of Rational Equations

Read the list of "THINGS TO KNOW" and review any concepts you are unfamiliar with.

Topic 16.7 Objective 1: Use Proportions to Solve Problems

What is a **proportion**?

Example 1:
Study the solution for Example 1 on page 16.7-3 and record the answer below.

A quality-control inspector examined a sample of 200 light bulbs and found 18 of them to be defective. At this ratio, how many defective bulbs can the inspector expect in a shipment of 22,000 light bulbs?

Example 2:
Complete Example 2 on page 16.7-5 on your own. Check your answer by clicking on the link. If your answer is incorrect, watch the video to find your error.

A landscaper plants grass seed at a general rate of 7 pounds for every 1000 square feet. If the landscaper has 25 pounds of grass seed on hand, how many additional pounds of grass seed will he need to purchase for a job to plant grass o a 45,000 square-foot yard?

What are **similar triangles**?

In similar triangles corresponding angles are _____.

In similar triangles corresponding sides are _____.

Topic 16.7

Example 3:

Study the solution for Example 3 on page 16.7-6 and record the answer below.

Find the unknown length for *n* for the following similar triangles.

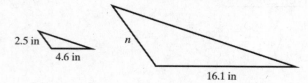

Example 4:

Complete Example 4 on page 16.7-8 on your own. Check your answer by clicking on the link. If your answer is incorrect, watch the video to find your error.

A forest ranger wants to determine the height of a tree. She measures the tree's shadow as 84 feet long. Her own shadow at the same time is 7.5 feet long. If she is 5.5 feet tall, how tall is the tree?

Topic 16.7 Objective 2: Use Formulas Containing Rational Expressions to Solve Problems

Example 5:

Study the solution for Example 5 on page 16.7-9 and record the answer below.

In electronics, the total resistance R of a circuit containing two resistors in parallel is given by the formula $\frac{1}{R} = \frac{1}{R_1} + \frac{1}{R_2}$, where R_1 and R_2 are the two individual resistances. If the total resistance is 10 ohms and one resistor has twice the resistance of the other, find the resistance of each circuit.

Topic 16.7 Objective 3: Solve Uniform Motion Problems Involving Rational Equations

Example 6:
Complete Example 6 on page 16.7-11 on your own. Check your answer by clicking on the link. If your answer is incorrect, watch the video to find your error.

Emalie can travel 16 miles upriver in the same amount of time it takes her to travel 24 miles downriver. If the speed of the current is 4 mph, how fast can her boat travel in still water?

Example 7:
Complete Example 7 on page 16.7-13 on your own. Check your answer by clicking on the link. If your answer is incorrect, watch the video to find your error.

Fatima rode an express train 223.6 miles from Boston to New York City and then rode a passenger train 218.4 miles from New York City to Washington, D.C. If the express train travels 30 miles per hour faster than the passenger train and her total trip took 6.5 hours, what was the average speed of the express train?

Topic 16.7 Objective 4: Solve Problems Involving Rate of Work

What is a **Rate of Work**?

Can rates be added?

Can times be added?

Topic 16.7

Example 8:
Study the solution for Example 8 on page 16.7-16 and record the answer below.

Avril can paint a room in 4 hours if she works alone. Anisa can paint the same room in 2 hours if she works alone. How long will it take the two women to paint the room if they work togeher?

Example 9:
Complete Example 9 on page 16.7-18 on your own. Check your answer by clicking on the link. If your answer is incorrect, watch the video to find your error.

A small pump takes 8 more hours than a larger pump to empty a pool. Together, the pumps can empty the pool in 3 hours. How long will it take the larger pump to empty the pool if it works alone?

Example 10:
Complete Example 10 on page 16.7-19 on your own. Check your answer by clicking on the link. If your answer is incorrect, watch the video to find your error.

A garden hose can fill a pond in 2 hours whereas an outlet pipe can drain the pond in 10 hours. If the outlet pipe is accidentally left open, how long would it take to fill the pond?

Topic 16.8 Guided Notebook

Topic 16.8 Variation

Read the list of "THINGS TO KNOW" and review any concepts you are unfamiliar with.

Topic 16.8 Objective 1: Solve Problems Involving Direct Variation

What is the purpose of **variation equation**?

Write down the definition of **Direct Variation**.

Example 1:
Study the solutions for Example 1 parts a and b on page 16.8-3 and record the answers below.

Suppose y varies directly with x, and $y = 20$ when $x = 8$.

a. Find the equation that relates x and y.

b. Find y when $x = 12$.

Example 2:
Complete Example 2 on page 16.8-5 on your own. Check your answer by clicking on the link. If your answer is incorrect, watch the video to find your error.

Suppose y varies directly with the cube of x, and $y = 375$ when $x = 5$.

a. Find the equation that relates x and y.

b. Find y when $x = 2$.

Topic 16.8

What is the four-step process for **Solving Variation Problems**?

1.

2.

3.

4.

Example 3:
Study the solution for Example 3 on page 16.8-6 and record the answer below.

The kinetic energy of an object in motion varies directly with the square of its speed. If a van traveling at the speed of 30 meters per second has 945,000 joules of kinetic energy, how much kinetic energy does it have it if is traveling at a speed of 20 meters per second?

Example 4:
Complete Example 4 on page 16.8-7 on your own. Check your answer by clicking on the link. If your answer is incorrect, watch the video to find your error.

The Ponderal Index measure of leanness states that weight varies directly with the cube of height. If a "normal" person who is 1.2 m tall weighs 21.6 kg, how much will a "normal" person weigh if they are 1.8 m tall?

Topic 16.8 Objective 2: Solve Problems Involving Inverse Variation

What is the definition of **Inverse Variation**?

Example 5:
Study the solution for Example 5 part a on page 16.8-10 and record the answer below. Complete part b on your own and check your answer by clicking on the link. If your answer is incorrect, watch the video to find your error.

Suppose y varies inversely with x, and $y = 72$ when $x = 50$.

a. Find the equation that relates x and y.

b. Find y when $x = 45$.

Example 6:
Study the solution for Example 6 on page 16.8-11 and record the answer below.

For a given mass, the density of an object is inversely proportional to its volume. If 50 cubic centimeters (cm^3), of an object with density of $28 g/cm^3$ is compressed to 40 cm^3, what would be its new density?

Topic 16.8

Example 7:

Complete Example 7 on page 16.8-12 on your own. Check your answer by clicking on the link. If your answer is incorrect, watch the video to find your error.

The shutter speed, S, of a camera varies inversely as the square of the aperture setting, f. if the shutter speed is 125 for an aperture of 5.6, what is the shtter speed if the aperture is 1.4?

Topic 17.1

Topic 17.1 Guided Notebook

Topic 17.1 Relations and Functions

Read the list of "THINGS TO KNOW" and review any concepts you are unfamiliar with.

Topic 17.1 Objective 1: Identify Independent and Dependent Variables

When is a variable **dependent**?

When is a variable **independent**?

Example 1:
Study the solutions for Example 1 parts a – c on page 17.1-4 and record the answers below.

For each of the following equations, identify the dependent variable and the independent variable(s).

a. $y = 3x + 5$
b. $w = ab + 3c^2$
c. $3x^2 + 9y = 12$

Topic 17.1 Objective 2: Find the Domain and Range of a Relation

What is a **relation**?

What is the **domain**?

What is the **range**?

Topic 17.1

Example 2:
Complete Example 2 on page 17.1-6 on your own. Check your answer by clicking on the link. If your answer is incorrect, watch the video to find your error.

Find the domain and range of each relation.

a. $\{(-5,7),(3,5),(6,7),(12,-4)\}$

b.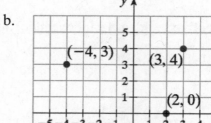

Example 3:
Complete Example 3 on page 17.1-7 on your own. Check your answer by clicking on the link. If your answer is incorrect, watch the video to find your error.

a.   b.   c.

What is a **feasible domain**?

Topic 17.1 Objective 3: Determine If Relations are Functions

Define a **Function**.

How do you determine if a set of ordered pairs is a function?

406

Topic 17.1

How do you determine if an equation is a function?

Example 4:
Complete Example 4 parts a – f on page 17.1-10 on your own. Check your answers by clicking on the link. If your answers are incorrect, watch the video to find your error.

Determine if each of the following relations is a function.

a. $\{(-3,6),(2,5),(0,6),(17,-9)\}$

b. $\{(4,5),(7,-3),(4,10),(-6,1)\}$

c. $\{(-2,3),(0,3),(4,3),(6,3),(8,3)\}$

d. $|y-5| = x+3$

e. $y = x^2 - 3x + 2$

f. $4x - 8y = 24$

Topic 17.1 Objective 4: Determine If Graphs Are Functions

What is the **Vertical Line Test** and what is it used for? Watch the animation on 17.1-11.

Example 5:
Complete Example 5 parts a – f on pages 17.1-11 and 17.1-12 on your own. Check your answers by clicking on the link. If your answers are incorrect, watch the video to find your error.

a.   b.   c.

d.   e.   f.

Topic 17.1

Topic 17.1 Objective 5: Solve Application Problems Involving Relations and Functions

Example 6:
Study the solutions for Example 6 parts a – d on page 17.1-13 and record the answers below.

The data in the table on page 17.1-13 represent the average daily hours of sleep and average daily hours of video entertainment for six students at a local college.

a. If a researcher believes the number of hours of video entertainment affects the number of hours of sleep, identify the independent variable and the dependent variable.

b. What are the ordered pairs for this data?

c. What are the domain and range?

d. Is this relation a function? Explain.

Example 7:
Study the solution for Example 7 part a on page 17.1-15 and record the answer below. Complete parts b - d on your own and check your answers by clicking on the link. If your answers are incorrect, watch the video to find your error.

The percent of households, y, with high-speed internet access in 2007 can be modeled by the equation $y = 0.70x + 20.03$, where x is the annual household income (in $1000s). (*Source:* U.S. Department of Commerce)

a. Identify the independent and dependent variables.

b. Use the model equation to estimate the percent of households in 2007 with high-speed internet access (to the nearest whole percent) if the annual household income was $50,000. What point would this correspond to on the graph of the equation?

c. Is the relation a function? Explain

d. Determine the feasible domain.

Topic 17.2 Guided Notebook

Topic 17.2 Function Notation and the Algebra of Functions

Read the list of "THINGS TO KNOW" and review any concepts you are unfamiliar with.

Topic 17.2 Objective 1: Express Equations of Functions Using Function Notation

What is an example of **function notation** and how is it read?

Read and summarize the CAUTION statement on page 17.2-3

What is a benefit of function notation?

Record the steps for **Expressing Equations of Functions Using Function Notation**.

1.

2.

3.

Read and summarize the CAUTION statement on 17.2-4

Topic 17.2

Example 1:
Study the solutions for Example 1 parts a and b on page 17.2-5, and record the answers below. Complete part c on your own and check your answer by clicking on the link. If your answer is incorrect, watch the video to find your error.

Write each function using function notation. Let x be the independent variable and y be the dependent variable.

a. $y = 2x^2 - 4$
b. $y - \sqrt{x} = 0$
c. $3x + 2y = 6$

Topic 17.2 Objective 2: Evaluate Functions

What does $f(x)$ represent and what is it called?

How do you **evaluate a function**?

Example 2:
Study the solution for Example 2 part a on page 17.2-6, and record the answer below. Complete parts b – d on your own and check your answers by clicking on the link. If your answers are incorrect, watch the video to find your error.

If $f(x) = 4x - 5$, $g(t) = 3t^2 - 2t + 1$ and $h(r) = \sqrt{r} - 9$, evaluate each of the following.

a. $f(3)$
b. $g(-1)$
c. $h(16)$
d. $f\left(\dfrac{1}{2}\right)$

Example 3:
Study the solution for Example 3 part a on page 17.2-8, and record the answer below. Complete parts b – d on your own and check your answers by clicking on the link. If your answers are incorrect, watch the video to find your error.

If $P(x) = 4x^3 - 2x^2 + 8x + 7$, evaluate each of the following.

a. $P(4)$
b. $P(-2)$
c. $P\left(-\dfrac{1}{2}\right)$

Example 4:
Study the solution for Example 4 part a on page 17.2-9, and record the answer below. Complete part b on your own and check your answer by clicking on the link. If your answer is incorrect, watch the video to find your error.

If $R(x) = \dfrac{5x^2 - 9}{7x + 3}$, evaluate each of the following.

a. $R(1)$

b. $R(-3)$

Topic 17.2 Objective 3: Find the Domain of a Polynomial or Rational Function

What is the **domain** of a polynomial function?

What is the **domain** of a rational function?

What is the procedure for **Finding the Domain of a Rational Function**?

Example 5:
Study the solutions for Example 5 on page 17.2-12.

Example 6:
Complete Example 6 on page 17.2-13 on your own. Check your answer by clicking on the link. If your answer is incorrect, watch the video to find your error.

Find the domain of $g(x) = \dfrac{x^2 + 2x - 15}{x^2 + 5x - 24}$

Topic 17.2

Topic 17.2 Objective 4: Find the Sum, Difference, Product, and Quotient of Functions

Record the **Algebra of Functions**.

1.

2.

3.

4.

Example 8:
Study the solution for Example 8 part a on page 17.2-16 and record the answer below. Complete part b on your own and check your answer by clicking on the link. If your answer is incorrect, watch the video to find your error.

For $P(x) = x^4 - 9x^2 + 7$ and $Q(x) = 3x^4 - 4x^2 + 2x - 10$, find each of the following.

a. $(P+Q)(x)$

b. $(P-Q)(x)$

Example 10:
Complete Example 10 on page 17.2-18 on your own. Check your answer by clicking on the link. If your answer is incorrect, watch the video to find your error.

For $P(x) = 15x^3 + 41x^2 + 4x + 3$ and $Q(x) = 5x + 2$, find $\left(\dfrac{P}{Q}\right)(x)$. State any values that cannot be included in the domain of $\left(\dfrac{P}{Q}\right)(x)$. (Note that $Q(x)$ cannot be 0.)

Topic 17.3

Topic 17.3 Guided Notebook

Topic 17.3 Graphs of Functions and Their Applications

Read the list of "THINGS TO KNOW" and review any concepts you are unfamiliar with.

Topic 17.3 Objective 1: Graph Simple Functions by Plotting Points

What is the **Graph of a Function**?

Write down the definition of a **Linear Function**.

Record the **Strategy for Graphing Simple Functions by Plotting Points**.

1.

2.

3.

Example 1:
Complete Example 1 parts a – c on page 17.3-4 on your own. Check your answers by clicking on the link. If your answers are incorrect, watch the video to find your error.

Graph each function by plotting points.
a. $f(x) = 2x - 1$
b. $g(x) = x^2 + 2x - 3$
c. $h(x) = 2|x| - 1$

Topic 17.3

Topic 17.3 Objective 2: Interpret Graphs of Functions

Example 2:
Study the solutions for Example 2 on page 17.3-7.

Example 3:
Complete Example 3 parts a – e on page 17.3-9 on your own. Check your answers by watching the video.

The graph of the function in Figure 2 gives the outside temperatures over one 24-hour period in spring. Use the given graph to answer the questions.

a. Over what time periods was the temperature rising?

b. Over what time periods was the temperature falling?

c. What was the highest temperature for the day? At what time was it reached?

d. What was the lowest temperature for the day shown? At what time was it reached?

e. Over what time period did the temperature decrease most rapidly?

Example 4:
Complete Example 4 on page 17.3-10 on your own. Check your answer by clicking on the link. If your answer is incorrect, watch the video to find your error.

A Boeing 757 jet took off and climbed steadily for 20 minutes until it reached an altitude of 18,000 feet. The jet maintained that altitude for 30 minutes. Then it climbed steadily for 10 minutes until it reached an altitude of 26,000 feet. The jet remained at 26,000 feet for 40 minutes. Then it descended steadily for 20 minutes until it reached an altitude of 20,000 feet, where it remained for 30 more minutes. During the final 20 minutes of the flight, the jet descended steadily until it landed at its destination airport. Draw a graph of the 757's altitude as a function of time.

Topic 17.3 Objective 3: Solve Application Problems Involving Functions

Example 5:
Study the solutions for Example 5 on page 17.3-11 and record the answers below (see video for part f).

A rock is dropped from the top of a cliff. Its height, h, above the ground, in feet, at t seconds is given by the function $h(t) = -16t^2 + 900$. Use the model to answer the following questions.

a. Evaluate $h(0)$. What does this value represent?

b. Evaluate $h(2)$. What does this value represent? How far has the rock fallen at this time?

c. Evaluate $h(10)$. Is this possible? Explain.

d. Evaluate $h(7.5)$. Interpret this result.

e. Determine the feasible domain and the range that makes sense (or feasible range) within the context of the problem.

f. Graph the function.

Topic 17.3

Example 6:
Study the solutions for Example 6 on page 17.3-13 and record the answers below (see video for part d).

The average monthly rent, R, for apartments in Queens, New York, is modeled by the function $R(a) = 2.2a$, where a is the floor area of the apartment in square feet. Use the model to answer the following questions.

a. What is the average monthly rent for apartments in Queens, New York, with a floor area of 800 square feet?

b. What is the floor area of an apartment if its rent if $1430 per month?

c. Determine the feasible domain and the feasible range of the function.

d. Graph the function.

Topic 18.1

Topic 18.1 Guided Notebook

Topic 18.1 Radical Expressions

Read the list of "THINGS TO KNOW" and review any concepts you are unfamiliar with.

Topic 18.1 Objective 1: Find Square Roots of Perfect Squares

What is the definition of **Principal and Negative Square Roots**?

Write a radical expression and label all its parts.

When will the square root simplify to a rational number?

Example 1:
Study the solutions for Example 1 parts a – c on page 18.1-4, and record the answers below. Complete parts d - f on your own and check your answers by clicking on the link. If your answers are incorrect watch the video to find your error.

Evaluate.

a. $\sqrt{64}$

b. $-\sqrt{169}$

c. $\sqrt{-100}$

d. $\sqrt{\dfrac{9}{25}}$

e. $\sqrt{0.81}$

f. $\sqrt{0}$

Read and summarize the CAUTION statement on 18.1-5.

417

Topic 18.1

Topic 18.1 Objective 2: Approximate Square Roots

What happens when the radicand is not a perfect square?

The principal square root of 12 should be between what two numbers? Explain why.

Example 2:
Study the solutions for Example 2 parts a – c on page 18.1-7, and record the answers below.

Use your calculator to approximate each square root and round the answer to three decimal places. Check that the answer is reasonable.

a. $\sqrt{5}$ b. $\sqrt{45}$ c. $\sqrt{103}$

Topic 18.1 Objective 3: Simplify Radical Expressions of the Form: $\sqrt{a^2}$

Does $\sqrt{a^2}$ always equal a? Explain.

Example 3:
Study the solutions for Example 3 parts a – c on page 18.1-9, and record the answers below. Complete parts d - f on your own and check your answers by clicking on the link. If your answers are incorrect watch the video to find your error.

Simplify.

a. $\sqrt{(-12)^2}$ b. $\sqrt{(2x-5)^2}$ c. $\sqrt{100x^2}$

d. $\sqrt{x^2+12x+36}$ e. $\sqrt{9x^4}$ f. $\sqrt{y^6}$

Topic 18.1 Objective 4: Find Cube Roots

What is the definition of **Cube Roots**?

Can cube roots have negative numbers in the radicand? Why or why not?

Is absolute value used when simplifying cube roots? Explain.

Example 4:
Study the solutions for Example 4 parts a – c on page 18.1-11. Complete parts d - f on your own and check your answers by clicking on the link. If your answers are incorrect watch the video to find your error.

Simplify.

d. $\sqrt[3]{0.064}$

e. $\sqrt[3]{\dfrac{8}{27}}$

f. $\sqrt[3]{-64y^9}$

Read and summarize the CAUTION statement on 18.1-11.

Topic 18.1 Objective 5: Find and Approximate *n*th Roots.

What is the definition of **Principal *n*th Roots**?

Topic 18.1

What is the **index** of the radical expression and what does it indicate?

Write down the technique for **Simplifying Radical Expressions of the Form:** $\sqrt[n]{a^n}$

Example 5:
Study the solutions for Example 5 parts a – c on page 18.1-13. Complete parts d - f on your own and check your answers by clicking on the link. If your answers are incorrect watch the video to find your error.

Simplify.

d. $\sqrt[5]{x^{15}}$ 　　　　　　　　　　　e. $\sqrt[6]{(x-7)^6}$ 　　　　　　　　　　　f. $\sqrt[4]{-1}$

Read and summarize the CAUTION statement on 18.1-13.

Example 6:
Study the solutions for Example 6 parts a – c on page 18.1-14, and record the answers below.

Use your calculator to approximate each root and round the answer to three decimal places. Check that the answer is reasonable.

a. $\sqrt[3]{6}$ 　　　　　　　　　　　b. $\sqrt[4]{200}$ 　　　　　　　　　　　c. $\sqrt[5]{154}$

Topic 18.2 Guided Notebook

Topic 18.2 Radical Functions

Read the list of "THINGS TO KNOW" and review any concepts you are unfamiliar with.

Topic 18.2 Objective 1: Evaluate Radical Functions

What is a **radical function**?

What is the technique for evaluating functions as covered in Topic 17.2?

Example 1:
Study the solutions for Example 1 parts a – c on page 18.2-3, and record the answers below. Complete parts d - f on your own and check your answers by clicking on the link. If your answers are incorrect watch the video to find your error.

For the radical functions $f(x) = \sqrt{2x-5}$, $g(x) = \sqrt[3]{5x+9}$, and $h(x) = -3\sqrt[4]{x} + 2$ evaluate the following.

a. $f(15)$ b. $g(-2)$ c. $h(625)$

d. $g(-\frac{1}{5})$ e. $f(0.5)$ f. $h(1)$

Topic 18.2 Objective 2: Find the Domain of a Radical Function.

When can the radicand of a radical expression be negative and when must it be non-negative?

Topic 18.2

Record the **Guideline to Finding the Domain of a Radical Function**.

For an *even* index:

For an *odd* index:

Example 2:
Study the solution for Example 2 part a on page 18.2-6, and record the answer below. Complete parts b - c on your own and check your answers by clicking on the link. If your answers are incorrect watch the video to find your error.

Find the domain for each radical function.

a. $F(x) = \sqrt[4]{12 - 4x}$
b. $h(x) = \sqrt[5]{3x + 5}$
c. $G(x) = \sqrt[6]{5x + 7}$

Topic 18.2 Objective 3: Graph Functions That Contain Square Roots or Cube Roots

What is the **square root function**?

Complete the following chart to find ordered pairs that belong to the function.

x	$y = f(x) = \sqrt{x}$	(x, y)
0		
1		
2		
4		
6		
9		

Topic 18.2

Draw a sketch of the graph showing the ordered pairs of the points plotted.

Example 3:
Study the solution for Example 3 part a on page 18.2-8, and record the answer below. Complete parts b - c on your own and check your answers by clicking on the link. If your answers are incorrect watch the video to find your error.

Graph each function. Compare each graph to that of the square root function.

a. $F(x) = \sqrt{x+1}$
b. $g(x) = \sqrt{x} + 1$
c. $f(x) = -\sqrt{x}$

In part a, why were the numbers -1, 0, 3, and 8 chosen for x while in part b the numbers 0, 1, 4, and 9 were chosen for x?

What is the **cube root function**?

Copyright © 2014 Pearson Education, Inc.

Topic 18.2

Complete the following chart to obtain ordered pairs that belong to the function.

x	$y = f(x) = \sqrt[3]{x}$	(x, y)
-8		
-1		
0		
1		
8		

Draw a sketch of the graph showing the ordered pairs of the points plotted.

Example 4:

Study the solution for Example 4 part a on page 18.2-11, and record the answer below. Complete parts b - c on your own and check your answers by clicking on the link. If your answers are incorrect watch the video to find your error.

Graph each function. Compare each graph to that of the cube root function.

a. $F(x) = \sqrt[3]{x-2}$
b. $g(x) = \sqrt[3]{x} - 2$
c. $h(x) = -\sqrt[3]{x}$

Topic 18.3 Guided Notebook

Topic 18.3 Rational Exponents and Simplifying Radical Expressions

Read the list of "THINGS TO KNOW" and review any concepts you are unfamiliar with.

Topic 18.3 Objective 1: Use the Definition for Rational Exponents of the Form: $a^{\frac{1}{n}}$

What is the definition of a **Rational Exponent of the Form** $a^{\frac{1}{n}}$?

Example 1:
Study the solutions for Example 1 parts a – c page 18.3-4. Complete parts d and e on your own and check your answers by clicking on the link. If your answers are incorrect watch the video to find your error.

Write each exponential expression as a radical expression. Simplify if possible.

d. $(-81)^{\frac{1}{2}}$ e. $(7x^3y)^{\frac{1}{5}}$

Example 2:
Study the solution for Example 2 part a on page 18.3-5, and record the answer below. Complete parts b and c on your own and check your answers by clicking on the link. If your answers are incorrect watch the video to find your error.

Write each radical expression as an exponential expression.

a. $\sqrt{5y}$ b. $\sqrt[3]{7x^2y}$ c. $\sqrt[4]{\frac{2m}{3n}}$

Topic 18.3 Objective 2: Use the Definition for Rational Exponents of the Form: $a^{\frac{m}{n}}$

What is the **Definition of a Rational Exponent of the Form** $a^{\frac{m}{n}}$?

Topic 18.3

Example 3:

Study the solutions for Example 3 parts a – c on page 18.3-8. Complete parts d and e on your own and check your answers by clicking on the link. If your answers are incorrect watch the video to find your error.

Write each exponential expression as a radical expression. Simplify if possible.

d. $(-36)^{\frac{5}{2}}$

e. $(x^2 y)^{\frac{2}{5}}$

Example 4:

Study the solution for Example 4 part a on page 18.3-9, and record the answer below. Complete parts b and c on your own and check your answers by clicking on the link. If your answers are incorrect watch the video to find your error.

Write each radical expression as an exponential expression.

a. $\sqrt[8]{x^5}$

b. $\left(\sqrt[5]{2ab^2}\right)^3$

c. $\sqrt[4]{(10x)^3}$

Topic 18.3 Objective 3: Simplify Exponential Expressions Involving Rational Exponents

Review the **Rules for Exponents** found on page 18.3-12.

Example 6:

Study the solutions for Example 6 parts a – c on page 18.2-13. Complete parts d - f on your own and check your answers by clicking on the link. If your answers are incorrect watch the video to find your error.

Use the rules for exponents to simplify each expression. Assume all variables represent non-negative values.

d. $(32 x^{\frac{5}{6}} y^{\frac{10}{9}})^{\frac{3}{5}}$

e. $\left(\dfrac{125 x^{\frac{5}{4}}}{y^{\frac{7}{8}} z^{\frac{9}{4}}}\right)^{\frac{4}{3}}$

$y \neq 0, z \neq 0$

f. $\left(4 x^{\frac{1}{6}} y^{\frac{3}{4}}\right)^2 \left(3 x^{\frac{5}{9}} y^{-\frac{3}{2}}\right)$

$y \neq 0$

426

Topic 18.3

Topic 18.3 Objective 4: Use Rational Exponents to Simplify Radical Expressions

Write down the steps for **Using Rational Exponents to Simplify Radical Expressions:**

1.

2.

3.

Example 7:
Study the solutions for Example 7 parts a – c on page 18.3-16. Complete parts d - f on your own and check your answers by clicking on the link. If your answers are incorrect watch the video to find your error.

Use rational exponents to simplify each radical expression. Assume all variables represent non-negative values.

d. $\sqrt[8]{25x^2y^6}$ e. $\sqrt[4]{49}$ f. $\dfrac{\sqrt[3]{x}}{\sqrt[4]{x}}$
$x \neq 0$

Topic 18.3 Objective 5: Simplify Radical Expressions Using the Product Rule

What is the **Product Rule for Radicals**?

Read and summarize the CAUTION statement on page 18.3-18.

Record the steps for **Using the Product Rule to Simplify Radical Expressions of the Form** $\sqrt[n]{a}$

1.

2.

3.

Topic 18.3

Example 10:
Study the solution for Example 10 part a on page 18.3-23, and record the answer below. Complete parts b and c on your own and check your answers by clicking on the link. If your answers are incorrect watch the video to find your error.

Multiply and simplify. Assume all variables represent non-negative values.

a. $3\sqrt{10} \cdot 7\sqrt{2}$

b. $2\sqrt[3]{4} \cdot 5\sqrt[3]{6}$

c. $\sqrt[4]{18x^3} \cdot \sqrt[4]{45x^2}$

Topic 18.3 Objective 6: Simplify Radical Expressions Using the Quotient Rule

What is the **Quotient Rule for Radicals**?

Read and summarize the CAUTION statement on page 18.3-25.

What are the 3 conditions for a **Simplified Radical Expression**?

Condition 1

Condition 2

Condition 3

Example 12:
Study the solutions for Example 12 parts a and b on page 18.3-28. Complete parts c and d on your own and check your answers by clicking on the link. If your answers are incorrect watch the video to find your error.

Use the quotient rule to simplify. Assume all variables represent positive numbers.

c. $\dfrac{\sqrt{150m^9}}{\sqrt{3m}}$

d. $\dfrac{\sqrt{45x^5 y^{-3}}}{\sqrt{20xy^{-1}}}$

Topic 18.4

Topic 18.4 Guided Notebook

Topic 18.4 Operations with Radicals

Read the list of "THINGS TO KNOW" and review any concepts you are unfamiliar with.

Topic 18.4 Objective 1: Add and Subtract Radical Expressions

What is the definition of **Like Radicals**?

Give an example of like radicals and explain why they are like radicals.

Give an example of radicals that are not like radicals and explain why they are not like radicals.

Example 1:
Study the solutions for Example 1 parts a – c on page 18.4-5 and record the answers below.

Add or subtract.

a. $\sqrt{11} + 6\sqrt{11}$
b. $7\sqrt{3} - 5\sqrt[4]{3}$
c. $\sqrt[3]{\dfrac{5}{8}} + 2\sqrt[3]{5}$

Read and summarize the CAUTION statement on 18.4-8.

Example 3:
Study the solution for Example 3 part a on page 18.4-8, and record the answer below. Complete part b on your own and check your answer by clicking on the link. If your answer is incorrect watch the video to find your error.

Add or subtract.

a. $\sqrt{54} + 6\sqrt{72} - 3\sqrt{24}$
b. $\sqrt[3]{24} - \sqrt[3]{192} + 4\sqrt[3]{250}$

Topic 18.4

Example 5:

Complete Example 5 parts a – c on page 18.4-10 on your own. Check your answers by clicking on the link. If your answers are incorrect, watch the video to find your error.

Add or subtract. Assume variables represent non-negative values.

a. $\dfrac{\sqrt{45}}{6x} - \dfrac{4\sqrt{20}}{5x}$

b. $\dfrac{\sqrt[4]{a^5}}{3} + \dfrac{a\sqrt[4]{a}}{12}$

c. $\dfrac{3x^3\sqrt{24x^3y^3}}{2x\sqrt{3x^2y}} - \dfrac{x^2\sqrt{10xy^4}}{\sqrt{5y^2}}$

Topic 18.4 Objective 2: Multiply Radical Expressions

What property is used to multiply rational expressions? Illustrate with Example 6 part a on page 18.4-11.

Example 7:

Study the solution for Example 7 part a on page 18.4-13, and record the answer below. Complete part b on your own and check your answer by clicking on the link.

Multiply. Assume variables represent non-negative values.

a. $(7\sqrt{2} - 2\sqrt{3})(\sqrt{2} - 5)$

b. $(\sqrt{m} - 4)(3\sqrt{m} + 7)$

Give an example of conjugates that involve radicals.

If the product of conjugates involves square roots, what results?

Topic 18.4 Objective 3: Rationalize Denominators of Radical Expressions

What is the procedure to **Rationalize a Denominator with One Term?**

Example 9:
Study the solutions for Example 9 parts a and b on page 18.4-16 and record the answers below.

Rationalize the denominator.

a. $\dfrac{\sqrt{5}}{\sqrt{3}}$

b. $\sqrt{\dfrac{2}{5x}}$

Example 10:
Study the solution for Example 10 part a on page 18.4-17, and record the answer below. Complete part b on your own and check your answer by clicking on the link. If your answer is incorrect watch the video to find your error.

Rationalize the denominator.

a. $\sqrt[3]{\dfrac{11}{25x}}$

b. $\dfrac{\sqrt[4]{7x}}{\sqrt[4]{27y^2}}$

Read and summarize the CAUTION statement on 18.4-19.

Topic 18.4

Example 11:

Complete Example 11 parts a – c on page 18.4-19 on your own. Check your answers by clicking on the link. If your answers are incorrect, watch the video to find your error.

Simplify each expression first and then rationalize the denominator.

a. $\sqrt{\dfrac{3x}{50}}$

b. $\dfrac{\sqrt{18x}}{\sqrt{27xy}}$

c. $\sqrt[3]{\dfrac{-4x^5}{16y^5}}$

What is the procedure to **Rationalizing a Denominator with Two Terms**?

Example 12:

Study the solutions for Example 12 parts a and b on page 18.4-20, and record the answers below. Complete part c on your own and check your answer by clicking on the link. If your answer is incorrect watch the video to find your error.

Rationalize the denominator.

a. $\dfrac{2}{\sqrt{3}+5}$

b. $\dfrac{7}{3\sqrt{x}-4}$

c. $\dfrac{\sqrt{y}-3}{\sqrt{y}+2}$

Topic 18.5

Topic 18.5 Guided Notebook

Topic 18.5 Radical Equations and Models

Read the list of "THINGS TO KNOW" and review any concepts you are unfamiliar with.

Topic 18.5 Objective 1: Solve Equations Involving One Radical Expression

What is the definition of a **Radical Equation**? Provide two examples.

Read and summarize the CAUTION statement on 18.5-3.

What is the key to solving a radical equation?

How can the *isolated* radical be eliminated from an equation?

What are **extraneous solutions** and when can they occur? Can these solutions be included in the solution set?

Topic 18.5

Write down the steps for **Solving Equations Involving One Radical Expression**.

1.

2.

3.

4.

Read and summarize the CAUTION statement on 18.5-5

Example 1:
Study the solution for Example 1 on page 18.5-6.

Read and summarize the CAUTION statement on 18.5-7.

Example 3:
Complete Example 3 on page 18.5-8 on your own. Check your answer by clicking on the link. If your answer is incorrect, watch the video to find your error.

Solve $\sqrt{3x+7} - x = 1$.

Example 5:
Complete Example 5 on page 18.5-11 on your own. Check your answer by clicking on the link. If your answer is incorrect, watch the video to find your error.

Solve $(x^2-9)^{\frac{1}{4}}+3=5$.

Topic 18.5 Objective 2: Solve Equations Involving Two Radical Expressions

Write down the steps for **Solving Equations Involving Two Radical Expressions**.

1.

2.

3.

4.

Example 6:
Complete Example 6 on page 18.5-12 on your own. Check your answer by clicking on the link. If your answer is incorrect, watch the video to find your error.

Solve $\sqrt{x+9}-\sqrt{x}=1$

Topic 18.5

Example 7:
Complete Example 7 on page 18.5-14 on your own. Check your answer by clicking on the link. If your answer is incorrect, watch the video to find your error.

Solve $\sqrt{2x+3} + \sqrt{x-2} = 4$

Topic 18.5 Objective 3: Use Radical Equations and Models to Solve Application Problems

Example 9:
Study the solution for Example 9 part a on page 18.5-16. Complete part b on your own and check your answer by clicking on the link. If your answer is incorrect watch the video to find your error.

Solve each formula for the given variable.

b. Radius of a sphere: $r = \sqrt[3]{\dfrac{3V}{4\pi}}$ for V

Example 10:
Study the solution for Example 10 part a on page 18.5-18. Complete part b on your own and check your answer by clicking on the link. If your answer is incorrect watch the video to find your error.

A SMOG grade for written text is a minimum reading grade level G that a reader must possess in order to fully understand the written text being graded. If w is the number of words that have three or more syllables in a sample of 30 sentences from a given text, the SMOG grade for that text is given by the formula $G = \sqrt{w} + 3$. Use the SMOG grade formula to answer the following questions.

b. If a text must have a tenth-grade reading level, then how many words with three or more syllables would be needed in the sample of 30 sentences?

Topic 18.6 Guided Notebook

Topic 18.6 Complex Numbers

Read the list of "THINGS TO KNOW" and review any concepts you are unfamiliar with.

Topic 18.6 Objective 1: Simplify Powers of i

Why does the equation $x^2 + 1 = 0$ have no real solution? Find the answer by clicking on the popup.

Write down the definition of the **Imaginary Unit i**.

Complete the chart to simplify the powers of i:

i	
i^2	
i^3	
i^4	
i^5	
i^6	
i^7	
i^8	

Topic 18.6

What is the pattern for the powers of *i*?

Write down the steps for **Simplifying** i^n **for *n* > 4**.

1.

2.

3.

Example 1:
Study the solutions for Example 1 parts a and b on page 18.6-5, and record the answers below. Complete parts c - e on your own and check your answers by clicking on the link. If your answers are incorrect watch the video to find your error.

Simplify.

a. i^{17} b. i^{60} c. i^{39} d. $-i^{90}$ e. $i^{14} + i^{29}$

What is a **Complex Number**? What is the real part of a complex number? What is the imaginary part of a complex number?

Topic 18.6 Objective 2: Add or Subtract Complex Numbers

What is the technique for **Adding or Subtracting Complex Numbers**?

Example 2:
Study the solutions for Example 2 parts a and b on page 18.6-9. Complete part c on your own and check your answer by clicking on the link. If your answer is incorrect watch the video to find your error.

Perform the indicated operations.

c. $(-3-4i)+(2-i)-(3+7i)$

Topic 18.6 Objective 3: Multiply Complex Numbers

What is used to multiply complex numbers?

Example 4:
Complete Example 4 on page 18.6-12 on your own. Check your answer by clicking on the link. If your answer is incorrect, watch the video to find your error.

Multiply $(4-3i)(7+5i)$

Example 5:
Study the solution for Example 5 part a on page 18.6-13, and record the answer below. Complete part b on your own and check your answer by clicking on the link. If your answer is incorrect watch the video to find your error.

Multiply.

a. $(4+2i)^2$

b. $(\sqrt{3}-5i)^2$

Topic 18.6

Write down the definition of **Complex Conjugates**.

Topic 18.6 Objective 4: Divide Complex Numbers

Example 7:
Study the solution for Example 7 on page 18.6-16 and record the answer below.

Divide. Write the quotient in stand form. $\dfrac{1-3i}{5-2i}$

Topic 18.6 Objective 5: Simplify Radicals with Negative Radicands

What is the rule for the **Square Root of a Negative Number**?

Read and summarize the CAUTION statement on 18.6-20.

Example 10:
Study the solutions for Example 10 parts a and b on page 18.6-21, and record the answers below. Complete parts c - d on your own and check your answers by clicking on the link. If your answers are incorrect watch the video to find your error.

Simplify.

a. $\sqrt{-8} + \sqrt{-18}$

b. $\sqrt{-8} \cdot \sqrt{-18}$

c. $\dfrac{6 + \sqrt{(6)^2 - 4(2)(5)}}{2}$

d. $\dfrac{4 - \sqrt{-12}}{4}$

Topic 19.1 Guided Notebook

Topic 19.1 Solving Quadratic Equations

Read the list of "THINGS TO KNOW" and review any concepts you are unfamiliar with.

Topic 19.1 Objective 1: Solve Quadratic Equations Using The Square Root Property

What is a **quadratic equation**?

What is the **Square Root Property**?

Example 1:
Study the solutions for Example 1 parts a and b on page 19.1-4. See how to *check* your answers by clicking on the link.

What is the four-step process for **Solving Quadratic Equations Using the Square Root Property**?

  1.

  2.

  3.

  4.

Example 2:
Study the solutions for Example 2 parts a and b on page 19.1-6. Complete parts c and d on your own and check your answers by clicking on the link. If your answers are incorrect watch the video to find your error.

Solve

c. $(x-1)^2 = 9$ d. $2(x+1)^2 - 17 = 23$

Topic 19.1

Topic 19.1 Objective 2: Solve Quadratic Equations by Completing the Square

What does it mean to **complete the square**? How do you find the appropriate constant to add?

Example 3:
Study the solutions for Example 3 parts a and b on page 19.1-10. Complete part c on your own and check your answer by clicking on the link. If your answer is incorrect watch the video to find your error.

What number must be added to make the binomial a perfect square trinomial?

c. $x^2 - \dfrac{3}{2}x$

Write down the steps for **Solving** $ax^2 + bx + c = 0, a \neq 0$ **by Completing the Square**

    1.

    2.

    3.

    4.

    5.

Example 5:
Study the solution for Example 5 on page 19.1-13 and record the answer below.

Solve $2x^2 - 10x - 6 = 0$ by completing the square.

Topic 19.1

Example 6:
Complete Example 6 on page 19.1-14 on your own. Check your answer by clicking on the link. If your answer is incorrect, watch the video to find your error.

Solve $3x^2 - 18x + 19 = 0$ by completing the square.

Topic 19.1 Objective 3: Solve Quadratic Equations Using the Quadratic Formula

Watch the animation on page 19.1-15 to see the derivation of the quadratic formula.

Write down the **Quadratic Formula**.

Read and summarize the CAUTION statement on 19.1-15.

Example 8:
Complete Example 8 on page 19.1-17 on your own. Check your answer by clicking on the link. If your answer is incorrect, watch the video to find your error.

Solve $3x^2 + 2x - 2 = 0$ using the quadratic formula.

Example 10:
Complete Example 10 on page 19.1-20 on your own. Check your answer by clicking on the link. If your answer is incorrect, watch the video to find your error.

Solve $14x^2 - 5x = 5x^2 + 7x - 4$ using the quadratic formula.

Topic 19.1

Topic 19.1 Objective 4: Use the Discriminant to Determine the Number and Type of Solutions to a Quadratic Equation

What is the **Discriminant**?

Complete the statements below:

If D > 0, then

If D < 0, then

If D = 0, then

Topic 19.1 Objective 5: Solve Equations That Are Quadratic in Form

If an equation is **quadratic in form**, what is used to change it into a quadratic equation?

Example 12:
Study the solution for Example 12 part a on page 19.1-24, and record the answer below. Complete parts b - d on your own and check your answers by clicking on the link. If your answers are incorrect watch the video to find your error.

Solve each equation.

a. $2x^4 - 11x^2 + 12 = 0$

b. $\left(\dfrac{1}{x-2}\right)^2 + \dfrac{2}{x-2} - 15 = 0$

c. $x^{\frac{2}{3}} - 9x^{\frac{1}{3}} + 8 = 0$

d. $3x^{-2} - 5x^{-1} - 2 = 0$

Topic 19.2 Guided Notebook

Topic 19.2 Quadratic Functions and Their Graphs

Read the list of "THINGS TO KNOW" and review any concepts you are unfamiliar with.

Topic 19.2 Objective 1: Identify the Characteristics of a Quadratic Function from its Graph

What is the definition of a **Quadratic Function**? What is the shape of its graph?

View the animation on 19.2-8.

Example 2:
Study the solutions for Example 2 parts a – e on page 19.2-9 and record the answers below.

a. Vertex (include *minimum value* and *maximum value* in your description)

b. Axis of symmetry

c. y-intercept

d. x-intercept(s)

e. Domain and range

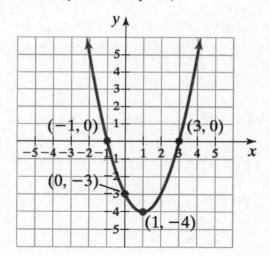

Topic 19.2 Objective 2: Graph Quadratic Functions by Using Translations

What is a **translation**?

What are **Vertical Shifts of Quadratic Functions**?

Topic 19.2

What are **Horizontal Shifts of Quadratic Functions**?

Example 5:
Study the solution for Example 5 on page 19.2-15.

Topic 19.2 Objective 3: Graph Quadratic Functions of the Form: $f(x) = a(x-h)^2 + k$

Define the **Standard Form of a Quadratic Function**.

What is another name for standard form?

Example 7:
Complete Example 7 on page 19.2-22 parts a – g on your own. Check your answers by clicking on the link. If your answers are incorrect, watch the video to find your error.

Given the quadratic function $f(x) = -(x-2)^2 - 4$, answer each of the following.

a. What are the coordinates of the vertex?

b. Does the graph "open up" or "open down"

c. What is the equation of the axis of symmetry?

d. Find any x-intercepts.

e. Find the y-intercept.

f. Sketch the graph.

g. State the domain and range in interval notation.

Topic 19.2 Objective 4: Find the Vertex of a Quadratic Function by Completing the Square

Write down the four steps for **Writing $f(x) = ax^2 + bx + c$ in Standard Form by Completing the Square**.

1.

2.

3.

4.

Topic 19.2 Objective 5: Graph Quadratic Functions of the Form $f(x) = ax^2 + bx + c$ by Completing the Square

Example 10:
Complete Example 10 on page 19.2-26 parts a – g on your own. Check your answers by clicking on the link. If your answers are incorrect, watch the video to find your error.

Rewrite the quadratic function $f(x) = 2x^2 - 4x - 3$ in standard form, and then answer the following:

a. What are the coordinates of the vertex?

b. Does the graph "open up" or "open down"

c. What is the equation of the axis of symmetry?

d. Find any x-intercepts.

e. Find the y-intercept.

f. Sketch the graph.

g. State the domain and range in interval notation.

Topic 19.2

Topic 19.2 Objective 6: Find the Vertex of a Quadratic Function by Using the Vertex Formula

Write down the **Formula for the Vertex of a Parabola**.

Example 11:
Study the solution for Example 11 part a on page 19.2-29. Complete part b on your own and check your answer by clicking on the link. If your answer is incorrect watch the video to find your error.

Use the vertex formula to find the vertex for each quadratic function.

b. $f(x) = -\frac{1}{2}x^2 - 10x + 5$

Topic 19.2 Objective 7: Graph Quadratic Functions of the Form $f(x) = ax^2 + bx + c$ by Using the Vertex Formula

Example 12:
Complete Example 12 on page 19.2-30 parts a – g on your own. Check your answers by clicking on the link. If your answers are incorrect, watch the video to find your error.

Given the quadratic function $f(x) = -2x^2 - 4x + 5$, answer the following:

a. What are the coordinates of the vertex?

b. Does the graph "open up" or "open down"

c. What is the equation of the axis of symmetry?

d. Find any x-intercepts.

e. Find the y-intercept.

f. Sketch the graph.

g. State the domain and range in interval notation.

Topic 19.3 Guided Notebook

Topic 19.3 Applications and Modeling of Quadratic Functions

Read the list of "THINGS TO KNOW" and review any concepts you are unfamiliar with.

Topic 19.3 Objective 1: Solve Applications Involving Unknown Numbers

Review the six steps of the **Problem-Solving Strategy for Applications**

Example 2:
Study the solutions for Example 2 on page 19.3-5 and record the answer below.

Three consecutive positive even integers are such that the square of the third is 20 less than the sum of the squares of the first two. Find the positive integers.

Topic 19.3 Objective 2: Solve Applications Involving Projectile Motion

Example 3:
Study the solutions for Example 3 on page 19.3-8 and record the answer below.

A toy rocket is launched at an initial velocity of 14.7 m/s from a platform that sits 49 meters above the ground. The height h of the rocket above the ground at any time t seconds after launch is given by the equation $h = -4.9t^2 + 14.7t + 49$. When will the rocket hit the ground?

Topic 19.3 Objective 3: Solve Applications Involving Geometric Formulas

Example 4:
Complete Example 4 on page 19.3-10 on your own. Check your answer. If your answer is incorrect, watch the video to find your error.

The length of a rectangle is 6 inches less than four times the width. Find the dimensions of the rectangle if the area of the rectangle is 54 square inches.

Topic 19.3

Topic 19.3 Objective 4: Solve Applications Involving Distance, Rate, and Time

Example 6:
Complete Example 6 on page 19.3-12 on your own. Check your answer by clicking on the link. If your answer is incorrect, watch the video to find your error.

Kevin flew his new Cessna O-2A airplane from Jonesburg to Mountainview, a distance of 2560 miles. The average speed for the return trip was 64 mph faster than the average outbound speed. If the total flying time for the round trip was 18 hours, what was the plane's average speed on the outbound trip from Jonesburg to Mountainview?

Topic 19.3 Objective 5: Solve Applications Involving Work

Example 7:
Complete Example 7 on page 19.3-14 on your own. Check your answer by clicking on the link. If your answer is incorrect, watch the video to find your error.

Dawn can finish the monthly sales reports in 2 hours less time than it takes Adam. Working together, they were able to finish the sales reports in 8 hours. How long does it take each person to finish the monthly sales reports alone? (Round to the nearest minute)

Topic 19.3 Objective 6: Maximize Quadratic Functions to Solve Application Problems

When does a quadratic function have a **minimum** value and where is it located?

When does a quadratic function have a **maximum** value and where is it located?

Example 8:
Study the solution for Example 8 on page 19.3-17.

Topic 19.3

What is **revenue**? How do you compute the **total revenue**?

Example 10:
Study the solutions for Example 10 parts a and b on page 19.3-22, and record the answers below. Complete parts c and d on your own and check your answers by clicking on the link. If your answers are incorrect watch the video to find your error.

To sell x waterproof CD alarm clocks, WaterTime, LLC, has determined that the price in dollars must be $p = 250 - 2x$, which is the demand equation. Each clock costs $2 to produce, with fixed costs of $4000 per month, producing the cost function $C(x) = 2x + 4000$.

a. Express the revenue R as a function of x.

b. Express the profit P as a function of x.

c. Find the value of x that maximizes profit. What is the maximum profit?

d. What is the price of the alarm clock that will maximize profit?

Topic 19.3

Topic 19.3 Objective 7: Minimize Quadratic Functions to Solve Application Problems

Example 12:
Study the solution for Example 12 on page 19.3-26.

Example 14:
Complete Example 14 parts a – c on page 19.3-29 on your own. Check your answers by clicking on the link. If your answers are incorrect, watch the video to find your error.

An account rep in one territory oversees $N = 20$ accounts and a second account rep in a nearby territory manages $N = 8$ accounts. The long run average cost function for their industry is $C = N^2 - 70N + 1400$.

a. Determine the long run average cost for $N = 20$ accounts and $N = 8$ accounts.

b. What number of accounts minimizes the long run average cost? What is the minimum long run average cost?

c. Should the two territories be merged into a single territory?

Topic 19.4

Topic 19.4 Guided Notebook

Topic 19.4 Circles

Read the list of "THINGS TO KNOW" and review any concepts you are unfamiliar with.

Topic 19.4 Objective 1: Find the Distance between Two Points

Watch the video on page 19.4-3 and write down the **Distance Formula**.

Read and summarize the CAUTION statement on 19.4-4.

Example 1:
Study the solution for Example 1 on page 19.4-5 and record the answer below.

Find the distance $d(A, B)$ between points (-1, 5) and (4, -5).

Topic 19.4 Objective 2: Find the Midpoint of a Line Segment

Write down the **Midpoint of a Line Segment**.

Topic 19.4

Read and summarize the CAUTION statement on 19.4-6.

Read and summarize the CAUTION statement on 19.4-7.

Example 2:
Study the solution for Example 2 on page 19.4-8 and record the answer below.

Find the midpoint of the line segment with endpoints (-3, 2) and (4, 6).

Topic 19.4 Objective 3: Write the Standard Form of an Equation of a Circle

Watch the animation on page 19.4-9 and show the steps to develop the **Standard Form of the Equation of a Circle** using the distance formula.

Topic 19.4

Example 5:
Complete Example 5 on page 19.4-12 on your own. Check your answer by clicking on the link. If your answer is incorrect, watch the video to find your error.

Write the standard form of the equation of the circle with center (0, -4) and radius $r = \sqrt{5}$.

Topic 19.4 Objective 4: Sketch the Graph of a Circle Given in Standard Form

Example 8:
Complete Example 8 on page 19.4-16 on your own. Check your answer by clicking on the link. If your answer is incorrect, watch the video to find your error.

Find the center and the radius, and sketch the graph of the circle $(x-1)^2 + (y+2)^2 = 9$. Also find any intercepts.

Topic 19.4 Objective 5: Write the General Form of a Circle in Standard Form and Sketch Its Graph

What is the **General Form of the Equation of a Circle**?

What procedure is used to change from the general form of a circle to the standard form of a circle?

Topic 19.4

Example 10:
Study the solution for Example 10 on page 19.4-23 and record the answer below.

Write the equation $x^2 + y^2 - 8x + 6y + 16 = 0$ in standard form; find the center, radius, and intercepts, and sketch the graph.

Example 11:
Watch the animation for Example 11 on page 19.4-25 and answer the question below.

Write the equation $4x^2 + 4y^2 + 4x - 8y + 1 = 0$ in standard form; find the center, radius, and intercepts, and sketch the graph.

Topic 19.5 Guided Notebook

Topic 19.5 Polynomial and Rational Inequalities

Read the list of "THINGS TO KNOW" and review any concepts you are unfamiliar with.

Topic 19.5 Objective 1: Solve Polynomial Inequalities

Define a **Polynomial Inequality**.

What plays an important part when solving polynomial inequalities? Study Figure 21 on page 19.5-4 to understand why.

Example 1:
Watch the video for Example 1 on page 19.5-5 and answer the question below. Record the seven steps for **Solving Polynomial Inequalities** as you work the problem.

Solve $x^3 - 3x^2 + 2x \geq 0$.

1.

2.

3.

4.

5.

6.

7.

Topic 19.5

Example 2:

Study the solution for Example 2 on page 19.5-9 and record the answer below.

Solve $x^2 + 5x < 3 - x^2$.

Topic 19.5 Objective 2: Solve Rational Inequalities

Define a **Rational Inequality**.

Write down the steps for **Solving Rational Inequalities.**

    1.

    2.

    3.

    4.

    5.

    6.

    7.

Topic 19.5

Example 3:
Study the solution for Example 3 on page 19.5-15 and record the answer below.

Solve $\dfrac{x-4}{x+1} \geq 0$.

Read and summarize the CAUTION statement on 19.5-18.

Example 4:
Complete Example 4 on page 19.5-18 on your own. Check your answer by clicking on the link. If your answer is incorrect, watch the video to find your error.

Solve $x > \dfrac{3}{x-2}$.

Topic 19.5

Example 5:
Complete Example 5 on page 19.5-20 on your own. Check your answer by clicking on the link. If your answer is incorrect, watch the video to find your error.

Solve $\dfrac{x+1}{x-2} > \dfrac{7x+1}{x^2+x-6}$

Topic 20.1 Guided Notebook

Topic 20.1 Transformations of Functions

Read the list of "THINGS TO KNOW" and review any concepts you are unfamiliar with.

Topic 20.1 Objective 1: Use Vertical Shifts to Graph Functions

Click on each of the **Basic Functions** to review the properties of each graph.

Example 1:
Study the solution for Example 1 on page 20.1-4 and record the answer below.

Sketch the graphs of $f(x) = |x|$ and $g(x) = |x| + 2$.

Write down the information for **Vertical Shifts of Functions**.

Topic 20.1 Objective 2: Use Horizontal Shifts to Graph Functions

Study the graphs of $f(x) = x^2$ and $g(x) = (x+2)^2$, found in Figure 2 on page 20.1-7.

Topic 20.1

Write down the information for **Horizontal Shifts of Functions**

Example 2:
Watch the animation for Example 2 on page 20.1-10 and sketch the graph below.

Use the graph of $y = x^3$ to sketch the graph of $g(x) = (x-1)^3 + 2$.

Topic 20.1 Objective 3: Use Reflections to Graph Functions

Write down the information about **Reflections of Functions about the x-Axis**. Watch the animation on page 20.1-12.

Write down the information about **Reflections of Functions about the y-Axis**. Watch the animation on page 20.1-14.

Topic 20.1

Topic 20.1 Objective 4: Use Vertical Stretches and Compressions to Graph Functions

Write down the information about **Vertical Stretches and Compressions of Functions**. Watch the animations on page 20.1-19.

Topic 20.1 Objective 5: Use Horizontal Stretches and Compressions to Graph Functions

Write down the information about **Horizontal Stretches and Compressions of Functions**. Watch the animations on page 20.1-20.

Topic 20.1 Objective 6: Use Combinations of Transformations to Graph Functions

Write down the "order of operations" for sketching a function that involves multiple transformations.

     1.

     2.

     3.

     4.

     5.

     6.

Topic 20.1

Example 6:
Watch the animation for Example 6 on page 20.1-24 and sketch the graph below.

Use transformations to sketch the graph of $f(x) = -2(x+3)^2 - 1$.

Example 7:
Study the solutions for Example 7 parts a – c on page 20.1-25 and record the answers below.

Draw the graph of $y = f(x)$ on page 20.1-25 and *label* the key ordered pairs.

Use the above graph of $y = f(x)$ to sketch each of the following functions.

a. $y = -f(2x)$

b. $y = 2f(x-3) - 1$

c. $y = -\dfrac{1}{2}f(2-x) + 3$

Topic 20.2 Guided Notebook

Topic 20.2 Composite and Inverse Functions

Read the list of "THINGS TO KNOW" and review any concepts you are unfamiliar with.

Topic 20.2 Objective 1: Form and Evaluate Composite Functions

Begin learning objective 1 by watching the video on 20.2-3.

Read and summarize the CAUTION statement on 20.2-4.

Example 1:
Study the solutions for Example 1 parts a – f on page 20.2-5. Watch the interactive video.

Let $f(x) = 4x + 1$, $g(x) = \dfrac{x}{x-2}$ and $h(x) = \sqrt{x+3}$.

a. Find the function $f \circ g$

b. Find the function $g \circ h$

c. Find the function $h \circ f \circ g$

d. Evaluate $(f \circ g)(4)$ or state that it is undefined.

e. Evaluate $(g \circ h)(1)$ or state that it is undefined.

f. Evaluate $(h \circ f \circ g)(6)$ or state that it is undefined.

Topic 20.2

Example 2:
Study the solutions for Example 2 parts a – e on page 20.2-7. Watch the interactive video.

Use the graph to evaluate each expression.

a. $(f \circ g)(4)$
b. $(g \circ f)(-3)$
c. $(f \circ f)(-1)$

d. $(g \circ g)(4)$
e. $(f \circ g \circ f)(1)$

Topic 20.2 Objective 2: Determine the Domain of Composite Functions

Example 3:
Study the solutions for Example 3 parts a and b on page 20.2-9. Watch the interactive video.

Let $f(x) = \dfrac{-10}{x-4}$ and $g(x) = \sqrt{5-x}$.

a. Find the domain of $f \circ g$

b. Find the domain of $g \circ f$

Topic 20.2 Objective 3: Determine If a Function is One-to-One Using the Horizontal Line Test

Begin learning objective 3 by watching the video on 20.2-11.

What is the definition of a **One-to-One Function**?

What is the **Horizontal Line Test**?

Topic 20.2

Example 4:
Complete Example 4 parts a – d on page 20.2-15 on your own. Check your answer by watching the animation.

Determine whether each function is one-to-one.

a. Use the given graph on 20.2-15

b. Use the given graph on 20.2-15

c. $f(x) = x^2 + 1$, $x \leq 0$

d. $f(x) = \begin{cases} 2x + 4 \text{ for } x \leq -1 \\ 2x - 6 \text{ for } x \geq 4 \end{cases}$

Topic 20.2 Objective 4: Verify Inverse Functions

What is the definition of an **Inverse Function**?

The _____ of f is exactly the same as the _____ of f^{-1}, and the _____ of f is exactly the same as the _____ of f^{-1}.

If the point (a, b) is on the graph of f then the point _____ is on the graph of f^{-1}.

Read and summarize the CAUTION statement on 20.2-18.

Example 5:
Complete Example 5 on page 20.2-20 on your own. If your answer is incorrect, watch the video to find your error.

Show that $f(x) = \dfrac{x}{2x+3}$ and $g(x) = \dfrac{3x}{1-2x}$ are inverse functions using the composition cancellation equations.

Topic 20.2

Topic 20.2 Objective 5: Sketch the Graphs of Inverse Functions

Given the graph of a one-to-one function, how can we obtain the graph of its inverse?

Example 6:
Study the solution for Example 6 on page 20.2-22. Watch the animation for more detail.

Sketch the graph of $f(x) = x^2 + 1$, $x \leq 0$ and its inverse. Also state the domain and range of f and f^{-1}.

Topic 20.2 Objective 6: Find the Inverse of a One-to-One Function

Write down the four steps to find the inverse of a one-to-one function.

1.

2.

3.

4.

Example 7:
Study the solution for Example 7 on page 20.2-25. Watch the animation and follow the four-step process.

Topic 20.3 Guided Notebook

Topic 20.3 Exponential Functions

Read the list of "THINGS TO KNOW" and review any concepts you are unfamiliar with.

Topic 20.3 Objective 1: Use the Characteristics of Exponential Functions

Write down the definition of an **Exponential Function**.

Complete Table 6 from page 20.3-4.

x	$y = 2^x$	$y = 3^x$	$y = \left(\dfrac{1}{2}\right)^x$	$y = \left(\dfrac{1}{3}\right)^x$
-2				
-1				
0				
1				
2				

What point do all four graphs have in common? Why is this?

Study the **Characteristics of Exponential Functions** on page 20.3-6.

Example 1:
Study the solution for Example 1 on page 20.3-7 and record the answer below.

Sketch the graph of $f(x) = \left(\dfrac{2}{3}\right)^x$

Topic 20.3

Example 2:
Study the solution for Example 2 on page 20.3-9, and record the answer below.

Find the exponential function $f(x) = b^x$ whose graph is given on 20.3-9.

Topic 20.3 Objective 2: Sketch the Graphs of Exponential Functions Using Transformations

Example 3:
Complete Example 3 on page 20.3-12 on your own. Check your answer by watching the video.

Use transformations to sketch the graph of $f(x) = -2^{x+1} + 3$.

How do you find the value of the y-intercept?

How do you find the value of the x-intercept?

Topic 20.3 Objective 3: Solve Exponential Equations by Relating the Bases

If the bases of an exponential equation are the same what can be said about the exponents?

Write down the **Method of Relating the Bases**.

Example 4:
Complete Example 4 parts a and b on page 20.3-15 on your own. If your answers are incorrect, watch the animation to find your error.

Solve the following equations.

a. $8 = \dfrac{1}{16^x}$

b. $\dfrac{1}{27^x} = \left(\sqrt[4]{3}\right)^{x-2}$

Topic 20.3 Objective 4: Solve Applications of Exponential Functions

Example 5:
Study the solutions for Example 5 on page 20.3-16.

Most golfers find that their golf skills improve dramatically at first and level off rather quickly. For Example, suppose that the distance (in yards) that a typical beginning golfer can hit a 3-wood after t weeks of practice on the driving range is given by the exponential function $d(t) = 225 - 100(2.7)^{-0.7t}$. This function has been developed after many years of gathering data on beginning golfers.

How far can a typical beginning golfer initially hit a 3-wood? How far can a typical beginning golfer hit a 3-wood after 1 week of practice on the driving range? After 5 weeks? After 9 weeks? Round to the nearest hundredth yard.

Topic 20.3

Write down the **Periodic Compound Interest Formula**. Be sure to state what each variable represents.

Example 6:
Study the solution for Example 6 on page 20.3-21.

Which investment will yield the most money after 25 years?

Investment A: $12,000 invested at 3% compounded monthly
Investment B: $10,000 invested at 3.9% compounded quarterly

Write down the **Present Value Formula**. Be sure to state what each variable represents.

Topic 20.4 Guided Notebook

Topic 20.4 The Natural Exponential Function

Read the list of "THINGS TO KNOW" and review any concepts you are unfamiliar with.

Topic 20.4 Objective 1: Use the Characteristics of the Natural Exponential Function

Begin learning objective 1 by watching the video on 20.4-3.

What is the definition of **Natural Base**?

Study the **Characteristics of the Natural Exponential Function**.

Example 1:
Study the solutions for Example 1 parts a – c on page 20.4-6, and record the answers below.

Evaluate each expression correctly to six decimal places.

a. e^2 b. $e^{-0.534}$ c. $1000e^{0.013}$

Topic 20.4 Objective 2: Sketch the Graphs of Natural Exponential Functions Using Transformations

Example 2:
Study the solutions for Example 2 on page 20.4-7, and record the answers below.

Use transformations to sketch the graph of $f(x) = -e^x + 2$. Determine the domain, range, and y-intercept and find the equation of any asymptotes.

Topic 20.4

Topic 20.4 Objective 3: Solve Natural Exponential Equations by Relating the Bases

Example 3:
Complete Example 3 parts a and b on page 20.4-9 on your own. Check your answers by watching the interactive video.

Use the method of relating the bases to solve each exponential equation.

a. $e^{3x-1} = \dfrac{1}{\sqrt{e}}$

b. $\dfrac{e^{x^2}}{e^{10}} = \left(e^x\right)^3$

Topic 20.4 Objective 4: Solve Applications of the Natural Exponential Function

Write down the **Continuous Compound Interest Formula**. Be sure to state what each variable represents.

Example 4:
Study the solutions for Example 4 on page 20.4-11, and record the answers below.

How much money would be in an account after 5 years if an original investment of $6000 was compounded continuously at 4.5%? Compare this amount to the same investment that was compounded daily. Round to the nearest cent.

Write down the **Present Value Formula**. Be sure to state what each variable represents.

Example 5:
Study the solution for Example 5 on page 20.4-13, and record the answer below.

Find the present value of $18,000 if interest is paid at a rate of 8% compounded continuously for 20 years. Round to the nearest cent.

Write down the model for **Exponential Growth.** Be sure to state what each variable represents.

Topic 20.4

Example 6:
Study the solutions for Example 6 parts a – c on page 20.4-15, and record the answers below.

The population of a small town follows the exponential growth model $P(t) = 900e^{0.015t}$, where t is the number of years after 1900.

Answer the following questions, rounding each answer to the nearest whole number:

a. What was the population of this town in 1900?

b. What was the population of this town in 1950?

c. Use this model to predict the population of this town in 2012.

Example 7:
Study the solutions for Example 7 parts a and b on page 20.4-16, and record the answers below.

Twenty years ago, the State of Idaho Fish and Game Department introduced a new breed of wolf into a certain Idaho forest. The current wolf population in this forest is now estimated at 825, with a relative growth rate of 12%.

Answer the following questions, rounding each answer to the nearest whole number:

a. How many wolves did the Idaho Fish and Game Department initially introduce into this forest?

b. How many wolves can be expected after another 20 years?

Topic 20.5 Guided Notebook

Topic 20.5 Logarithmic Functions

Read the list of "THINGS TO KNOW" and review any concepts you are unfamiliar with.

Topic 20.5 Objective 1: Use the Definition of a Logarithmic Function

Begin learning objective 1 by watching the video on 20.5-3.

Write down the definition of a **Logarithmic Function**.

Write down steps to find the inverse of $f(x) = b^x$.

1.

2.

3.

4.

Example 1:
Study the solutions to Example 1 parts a – c on page 20.5-5. Record the answers below.

Write each exponential equation as an equation involving a logarithm.

a. $2^3 = 8$
b. $5^{-2} = \dfrac{1}{25}$
c. $1.1^M = z$

Topic 20.5

Example 2:

Study the solutions to Example 2 parts a – c on page 20.5-7. Record the answers below.

Write each logarithmic equation as an equation involving an exponent.

a. $\log_3 81 = 4$ b. $\log_4 16 = y$ c. $\log_{3/5} x = 2$

Topic 20.5 Objective 2: Evaluate Logarithmic Expressions

What are two ways logarithms may be evaluated?

Example 3:

Study the solutions for Example 3 parts a – c on page 20.5-8, and record the answers below.

Evaluate each logarithm:

a. $\log_5 25$ b. $\log_3 \dfrac{1}{27}$ c. $\log_{\sqrt{2}} \dfrac{1}{4}$

Topic 20.5 Objective 3: Use the Properties of Logarithms

Write down the **General Properties of Logarithms**.

Write down the **Cancellation Properties of Exponentials and Logarithms**.

Topic 20.5

Example 4:
Study the solutions for Example 4 parts a – d on page 20.5-12, and record the answers below.

Use the properties of logarithms to evaluate each expression.

a. $\log_3 3^4$
b. $\log_{12} 12$
c. $7^{\log_7 13}$
d. $\log_8 1$

Topic 20.5 Objective 4: Use the Common and Natural Logarithms

What is the definition of a **common logarithm**?

What is the definition of a **natural logarithm**?

Example 5:
Study the solutions for Example 5 parts a – c on page 20.5-14, and record the answers below.

Write each exponential equation as an equation involving a common logarithm or natural logarithm.

a. $e^0 = 1$
b. $10^{-2} = \dfrac{1}{100}$
c. $e^K = w$

Example 6:
Work through Example 6 parts a – c on page 20.5-15 and record the answers below. If your answers are incorrect, watch the video to find your error.

Write each logarithmic equation as an equation involving an exponent.

a. $\log 10 = 1$
b. $\ln 20 = Z$
c. $\log(x-1) = T$

Topic 20.5

Example 7:
Study the solutions for Example 7 parts a – d on page 20.5-16, and record the answers below.

Evaluate each expression without the use of a calculator.

a. $\log 100$
b. $\ln \sqrt{e}$
c. $e^{\ln 51}$
d. $\log 1$

Topic 20.5 Objective 5: Use the Characteristics of Logarithmic Functions

Study the three steps used to sketch the graph of a logarithmic function.

Example 8:
Study the solution for Example 8 on page 20.5-18.

Topic 20.5 Objective 6: Sketch the Graphs of Logarithmic Functions Using Transformations

Example 9:
Study the solution for Example 9 on page 20.5-21.

Topic 20.5 Objective 7: Find the Domain of Logarithmic Functions

Find the domain of $f(x) = -\ln(x+2) - 1$

Example 10:
Study the solution for Example 10 on page 20.5-24, and record the answer below.

Find the domain of $f(x) = \log_5 \left(\dfrac{2x-1}{x+3} \right)$

Topic 20.6 Guided Notebook

Topic 20.6 Properties of Logarithms

Read the list of "THINGS TO KNOW" and review any concepts you are unfamiliar with.

Topic 20.6 Objective 1: Use the Product Rule, Quotient Rule, and Power Rule for Logarithms

Write down the **Properties of Logarithms**. Watch any *one* of the videos for a proof.

1.

2.

3.

Example 1:
Study the solutions for Example 1 parts a and b on page 20.6-4, and record the answers below.

Use the product rule for logarithms to expand each expression. Assume $x > 0$.

a. $\ln(5x)$

b. $\log_2(8x)$

Read and summarize the CAUTION statement on 20.6-4.

Example 2:
Study the solutions for Example 2 parts a and b on page 20.6-5, and record the answers below.

Use the quotient rule for logarithms to expand each expression. Assume $x > 0$.

a. $\log_5\left(\dfrac{12}{x}\right)$

b. $\ln\left(\dfrac{x}{e^5}\right)$

Topic 20.6

Read and summarize the CAUTION statement on 20.6-5

Example 3:
Study the solutions for Example 3 parts a and b on page 20.6-6, and record the answers below.

Use the power rule for logarithms to expand each expression. Assume $x > 0$.

a. $\log 6^3$

b. $\log_{1/2} \sqrt[4]{x}$

Read and summarize the CAUTION statement on 20.6-6.

Topic 20.6 Objective 2: Expand and Condense Logarithmic Expressions

Example 4:
Study the solutions for Example 4 parts a and b on page 20.6-7, and record the answers below.

Use properties of logarithms to expand the logarithmic expression as much as possible.

a. $\log_7 \left(49x^3 \sqrt[5]{y^2} \right)$

b. $\ln \left(\dfrac{(x^2 - 4)}{9e^{x^3}} \right)$

Example 5:
Study the solutions for Example 5 parts a and b on page 20.6-9, and record the answers below.

Use the properties of logarithms to rewrite the expression as a single logarithm.

a. $\dfrac{1}{2} \log(x-1) - 3\log z + \log 5$

b. $\dfrac{1}{3}(\log_3 x - 2\log_3 y) + \log_3 10$

482

Topic 20.6

Topic 20.6 Objective 3: Solve Logarithmic Equations Using the Logarithm Property of Equality

Write down the **Logarithm Property of Equality**.

Example 6:
Study the solutions for Example 6 parts a and b on page 20.6-12, and record the answers below.

Solve the following equations.

a. $\log_7(x-1) = \log_7 12$

b. $2\ln x = \ln 16$

Read and summarize the CAUTION statement on 20.6-13.

Topic 20.6 Objective 4: Use the Change of Base Formula

Write down the **Change of Base Formula**. Watch the video for the proof of the change of base formula.

Example 7:
Study the solutions for Example 7 parts a and b on page 20.6-15, and record the answers below.

Approximate the following expressions. Round each to four decimal places.

a. $\log_9 200$

b. $\log_{\sqrt{3}} \pi$

Topic 20.6

Example 8:
Study the solutions for Example 8 on page 20.6-16, and record the answer below.

Use the change of base formula and the properties of logarithms to rewrite as a single logarithm involving base 2.

$\log_4 x + 3\log_2 y$

Example 9:
Complete Example 9 on page 20.6-17 on your own. Watch the video to verify your answer.

Use the change of base formula and the properties of logarithms to solve the equation.

$2\log_3 x = \log_9 16$

Topic 20.7 Guided Notebook

Topic 20.7 Exponential and Logarithmic Equations

Read the list of "THINGS TO KNOW" and review any concepts you are unfamiliar with.

Topic 20.7 Objective 1: Solve Exponential Equations

Recall the **Logarithm property of equality**:

Recall the **Power rule for logarithms**:

Example 1:
Study the solution for Example 1 on page 20.7-5, and record the answer below.

Solve $2^{x+1} = 3$

Write down the procedure for **Solving Exponential Equations**.

Example 2:
Complete Example 2 parts a and b on page 20.7-6 on your own. Verify your answers. If your answers are incorrect, watch the video to find your error.

Solve each equation. For part b, round to four decimal places.

a. $3^{x-1} = \left(\dfrac{1}{27}\right)^{2x+1}$ b. $7^{x+3} = 4^{2-x}$

Topic 20.7

Example 3:
Study the solutions for Example 3 parts a and b on page 20.7-9, and record the answers below. Watch the video for further explanation, if needed.

Solve each equation. Round to four decimal places.

a. $25e^{x-5} = 17$

b. $e^{2x-1} \cdot e^{x+4} = 11$

Topic 20.7 Objective 2: Solve Logarithmic Equations

Recall the **Properties of Logarithms**:

1.

2.

3.

Example 4:
Study the solution for Example 4 on page 20.7-12, and record the answer below.

Solve $2\log_5(x-1) = \log_5 64$

Why must the solution $x = -7$ be discarded?

Topic 20.7

Read and summarize the CAUTION statement on 20.7-13.

Write down the steps for **Solving Logarithmic Equations**.

1.

2.

3.

4.

5.

Example 5:
Study the solution for Example 5 on page 20.7-14, and record the answer below.

Solve $\log_4(2x-1) = 2$

Topic 20.7

Example 6:
Study the solution for Example 6 on page 20.7-16, and record the answer below. Watch the video for further explanation, if needed.

Solve $\log_2(x+10) + \log_2(x+6) = 5$

Since both answers are negative, should they both be discarded? Explain.

Example 7:
Study the solution for Example 7 on page 20.7-17, and record the answer below.

Solve $\ln(x-4) - \ln(x-5) = 2$ Round to four decimal places.

Topic 20.8 Guided Notebook

Topic 20.8 Applications of Exponential and Logarithmic Functions

Read the list of "THINGS TO KNOW" and review any concepts you are unfamiliar with.

Topic 20.8 Objective 1: Solve Compound Interest Applications

Recall the **Periodic Compound Interest Formula** and the **Continuous Compound Interest Formula**. Be sure to state what each variable represents.

Example 1:
Study the solution for Example 1 on page 20.8-5, and record the answer below.

How long will it take (in years and months) for an investment to double if it earns 7.5% compounded monthly?

Example 2:
Study the solution for Example 2 on page 20.8-7, and record the answer below.

Suppose an investment of $5000 compounded continuously grew to an amount of $5130.50 in 6 months. Find the interest rate, and then determine how long it will take for the investment to grow to $6000. Round the interest rate to the nearest hundredth of a percent and the time to the nearest hundredth of a year.

Topic 20.8

Topic 20.8 Objective 2: Solve Exponential Growth and Decay Applications

Write down the **Exponential Growth** model. Be sure to state what each variable represents.

Example 3:
Study the solution for Example 3 on page 20.8-10, and record the answer below.

The population of a small town grows at a rate proportional to its current size. In 1900, the population was 900. In 1920, the population had grown to 1600. What was the population of this town in 1950? Round to the nearest whole number.

Write down the **Exponential Decay** model. Be sure to state what each variable represents.

Study the animation for **Half-Life** on page 20.8-13.

Example 4:
Study the solution for Example 4 on page 20.8-13, and record the answer below.

Suppose that a meteorite is found containing 4% of its original Krypton-99. If the half-life of Krypton-99 is 80 years, how old is the meteorite? Round to the nearest year.

Topic 20.8 Objective 3: Solve Logistic Growth Applications

Write down the **Logistic Growth** model. Be sure to state what each variable represents.

Example 5:
Study the solutions for Example 5 parts a – c on page 20.8-17, and record the answer below.

Ten goldfish were introduced into a small pond. Because of limited food, space, and oxygen, the carrying capacity of the pond is 400 goldfish. The goldfish population at any time t, in days, is modeled by the logistic growth function $F(t) = \dfrac{C}{1 + Be^{kt}}$. If 30 goldfish are in the pond after 20 days,

a. Find B.

b. Find k.

c. When will the pond contain 250 goldfish? Round to the nearest whole number.

Topic 20.8

Topic 20.8 Objective 4: Use Newton's Law of Cooling

Write down the model for **Newton's Law of Cooling**. Be sure to state what each variable represents.

Example 6:

Study the solution for Example 6 on page 20.8-21, and record the answer below.

Suppose that the temperature of a cup of hot tea obeys Newton's law of cooling. If the tea has a temperature of 200° F when it is initially poured and 1 minute later has cooled to 189° F in a room that maintains a constant temperature of 69° F, determine when the tea reaches a temperature of 146° F. Round to the nearest minute.

Topic 21.1 Guided Notebook

Introduction to Conics Topics
View the animations on each of the four conic Topics, making note of how each is formed.

Topic 21.1 The Parabola

Read the list of "THINGS TO KNOW" and review any concepts you are unfamiliar with.

Topic 21.1 Objective 1: <u>Work with the Equation of a Parabola with a Vertical Axis of Symmetry</u>

View the animation on page 21.1-3 and describe the **Characteristics of a Parabola**. Take detailed notes below.

In Topic 19.2 you studied quadratic functions and parabolas from the algebraic point of view. What is the **Geometric Definition of the Parabola**?

Topic 21.1

Record the **Equation of a Parabola in Standard Form with a Vertical Axis of Symmetry**.

Draw two sketches of a vertical parabola.

What is the vertex?

What is the distance from the vertex to the focus?

What is the distance from the vertex to the directrix?

What is the focus?

What is the equation of the directrix?

Example 1:
Study the solution for Example 1 on page 21.1-7, and record the answer below.

Find the vertex, focus and directrix of the parabola $x^2 = 8y$ and sketch its graph.

Topic 21.1

Example 2:
Study the solution for Example 2 on page 21.1-8, and record the answer below. Watch the video for further explanation, if needed.

Find the vertex, focus, and directrix of the parabola $-(x+1)^2 = 4(y-3)$ and sketch its graph.

Topic 21.1 Objective 2: Work with the Equation of a Parabola with a Horizontal Axis of Symmetry

Write down the **Equation of a Parabola in Standard Form with a Horizontal Axis of Symmetry**.

Draw two sketches of a horizontal parabola.

What is the vertex?

What is the distance from the vertex to the focus?

What is the distance from the vertex to the directrix?

What is the focus?

What is the equation of the directrix?

Topic 21.1

Example 3:
Study the solution for Example 3 on page 21.1-11.

Topic 21.1 Objective 3: Find the Equation of a Parabola Given Information about the Graph

Example 4:
Study the solution for Example 4 on page 21.1-12, and record the answer below. Watch the video for further explanation, if needed.

Find the standard form of the equation of the parabola with focus $\left(-3, \frac{5}{2}\right)$ and directrix $y = \frac{11}{2}$.

Example 5:
Study the solution for Example 5 on page 21.1-13.

Topic 21.1 Objective 4: Complete the Square to Find the Equation of a Parabola in Standard Form

Example 6:
Study the solution for Example 6 on page 21.1-15, and record the answer below.

Find the vertex, focus, and directrix and sketch the graph of the parabola $x^2 - 8x + 12y = -52$.

Topic 21.1 Objective 5: Solving Applications Involving Parabolas

Example 7:
Study the solution for Example 7 on page 21.1-18, and record the answer below.

Parabolic microphones can be seen on the sidelines of professional sporting events so that television networks can capture audio sounds from the players on the field. If the surface of a parabolic microphone is 27 centimeters deep and has a diameter of 72 centimeters at the top, where should the microphone be placed relative to the vertex of the parabola?

Topic 21.2

Topic 21.2 Guided Notebook

Topic 21.2 The Ellipse

Read the list of "THINGS TO KNOW" and review any concepts you are unfamiliar with.

Topic 21.2 Objective 1: Sketch the Graph of an Ellipse

What is the **Geometric Definition of the Ellipse?**

Watch the video on 21.2-5 describing horizontal and vertical ellipses. Draw a sketch and label the key features of each below.

Write down the **Equation of an Ellipse in Standard Form with Center (h,k) and a Horizontal Major Axis**.

What is the relationship between a and b?

What are the ordered pairs of the foci?

What are the order pairs of the vertices?

What are the endpoints of the major axis?

What is the relationship between a, b, and c?

Topic 21.2

Write down the **Equation of an Ellipse in Standard Form with Center (h,k) and a Vertical Major Axis**.

What is the relationship between a and b?

What are the ordered pairs of the foci?

What are the order pairs of the vertices?

What are the endpoints of the major axis?

What is the relationship between a, b, and c?

Example 1:
Study the solution for Example 1 on page 21.2-10, and record the answer below.

Sketch the graph of the ellipse $\dfrac{x^2}{25}+\dfrac{y^2}{4}=1$, and label the center, foci, and vertices.

Example 2:
Study the solution for Example 2 on page 21.2-11.

Topic 21.2 Objective 2: Find the Equation of an Ellipse Given Information about the Graph

Example 3:
Study the solution for Example 3 on page 21.2-12, and record the answer below.

Find the standard form of the equation of the ellipse with foci at (-6,1) and (-2,1) such that the length of the major axis is eight units.

Example 4:
Complete Example 4 on page 21.2-14 on your own. Check your answer by watching the video.

Determine the equation of the ellipse with foci located at (0,6) and (0,-6) that passes through the point (-5,6)

Topic 21.2 Objective 3: Complete the Square to Find the Equation of an Ellipse in Standard Form

Example 5:
Study the solution for Example 5 on page 21.2-15, and record the answer below. Find the center and foci and sketch the ellipse

$36x^2 + 20y^2 + 144x - 120y - 396 = 0$

Topic 21.2

Topic 21.2 Objective 4: Applications Involving Ellipses

Example 6:

Study the solution for Example 6 on page 21.2-19, and record the answer below.

A patient is placed in an elliptical tank that is 280 centimeters long and 250 centimeters wide to undergo sound wave lithotripsy treatment for kidney stones. Determine where the sound emitter and the stone should be positioned relative to the center of the ellipse.

Topic 21.3 Guided Notebook

Topic 21.3 The Hyperbola

Read the list of "THINGS TO KNOW" and review any concepts you are unfamiliar with.

Topic 21.3 Objective 1: Sketch the Graph of a Hyperbola

What is the **Geometric Definition of the Hyperbola**?

Draw a sketch of two different hyperbolas below.

Identify each of the following in your sketches above:
Center   Foci   Transverse axis
Vertices   Asymptotes   Conjugate Axis

Write down the **Equation of a Hyperbola in Standard Form with a Horizontal Transverse Axis**.

What are the ordered pairs of the foci?

What are the ordered pairs of the vertices?

What are the ordered pairs of the endpoints?

What is the relationship between a, b and c?

What are the equations of the asymptotes?

Topic 21.3

Write down the **Equation of a Hyperbola in Standard Form with a Vertical Transverse Axis**.

What are the ordered pairs of the foci?

What are the ordered pairs of the vertices?

What are the ordered pairs of the endpoints?

What is the relationship between a, b and c?

What are the equations of the asymptotes?

Write down the **Standard Equations of a Hyperbola with the Center at the Origin.**

Example 1:
Study the solutions for Example 1 parts a and b on page 21.3-12, and record the answer below.

Sketch the following hyperbolas. Determine the center, transverse axis, vertices, and foci and find the equations of the asymptotes.

a. $\dfrac{(y-4)^2}{36} - \dfrac{(x+5)^2}{9} = 1$

b. $25x^2 - 16y^2 = 400$

Topic 21.3 Objective 2: Find the Equation of a Hyperbola in Standard Form

Example 2:
Study the solution for Example 2 on page 21.3-14, and record the answer below.

Find the equation of the hyperbola with the center at (-1,0), a focus at (-11,0), and a vertex at (5,0).

Topic 21.3

Topic 21.3 Objective 3: Complete the Square to Find the Equation of a Hyperbola in Standard Form

Review how to **complete the square**.

Example 3:
Study the solution for Example 3 on page 21.3-16, and record the answer below.

Find the center, vertices, foci, and equations of asymptotes and sketch the hyperbola

$12x^2 - 4y^2 - 72x - 16y + 140 = 0$

Topic 21.3 Objective 4: Solve Applications Involving Hyperbolas

Example 4:
Study the solution for Example 4 on page 21.3-20, and record the answer below.

One transmitting station is located 100 miles due east from another transmitting station. Each station simultaneously sends out a radio signal. The signal from the west tower is received by a ship $\frac{1600}{3}$ microseconds after the signal from the east tower. If the radio signal travels at 0.18 miles per microsecond, find the equation of the hyperbola on which the ship is presently located.

Topic 22.1 Guided Notebook

Topic 22.1 Sequences and Series

Read the list of "THINGS TO KNOW" and review any concepts you are unfamiliar with.

Topic 22.1 Objective 1: Write the Terms of a Sequence

Write down the definition of a **Finite Sequence** and give one example.

Write down the definition of an **Infinite Sequence** and give one example.

Write down the definition of **The Factorial of a Non-Negative Integer**.

Example 1:
Study the solutions for Example 1 parts a – c on page 22.1-6 and record the answers below. Complete part d on your own and check your answer by clicking on the link. If your answer is incorrect, watch the video to find your error.

Write the first four terms of each sequence whose nth term is given.

a. $a_n = 2n - 1$ b. $b_n = n^2 - 1$ c. $c_n = \dfrac{3^n}{(n-1)!}$ d. $d_n = (-1)^n 2^{n-1}$

Topic 22.1 Objective 2 Write the Terms of a Recursive Sequence

Write down the definition of a **recursive sequence**.

Topic 22.1

Example 2:
Complete Example 2 parts a and b on page 22.1-8 on your own. If your answers are incorrect, watch the video to find your error.

Write the first four terms of each of the following recursive sequences.

a. $a_1 = -3, a_n = 5a_{n-1} - 1$ for $n \geq 2$

b. $b_1 = 2, b_n = \dfrac{(-1)^{n-1} n}{b_{n-1}}$ for $n \geq 2$

Example 3:
Study the solution for Example 3 on a page 22.1-9, and record the answer below.

The **Fibonacci sequence** is recursively defined by $a_n = a_{n-1} + a_{n-2}$ where $a_1 = 1$ and $a_2 = 1$. Write the first eight terms of the Fibonacci sequence.

Topic 22.1 Objective 3: Write the General Term for a Given Sequence

Example 4:
Study the solutions for Example 4 parts a and b on page 22.1-11, and record the answers below. Watch the video for further explanation, if needed.

Write a formula for the nth term of each infinite sequence then use this formula to find the 8th term of the sequence.

a. $\dfrac{1}{1}, \dfrac{1}{2}, \dfrac{1}{3}, \dfrac{1}{4}, \dfrac{1}{5}, \ldots$

b. $-\dfrac{2}{1}, \dfrac{4}{2}, -\dfrac{8}{6}, \dfrac{16}{24}, -\dfrac{32}{120}, \ldots$

Topic 22.1 Objective 4: Compute Partial Sums of a Series

Write down the definition of a **finite series**.

Write down the definition of an **infinite series**.

The sum of the first n terms of a series is called the _____ of the series and is denoted as _____.

Example 5:
Study the solutions for Example 5 parts a and b on page 22.1-15, and record the answers below.

Given the general term of each sequence, find the indicated partial term.

a. $a_n = \dfrac{1}{n}$ find S_3
b. $b_n = (-1)^n 2^{n-1}$ find S_5

Topic 22.1 Objective 5: Determine the Sum of a Finite Series Written in Summation Notation

Illustrate **Summation Notation** for both a finite series and an infinite series.

Using one of your Examples above, identify the **index of summation**, the **lower limit of summation**, and the **upper limit of summation**.

Example 6:

Study the solutions for Example 6 parts a – c on page 22.1-17, and record the answers below. Watch the video for further explanation, if needed.

Find the sum of each finite series.

a. $\displaystyle\sum_{i=1}^{5} i^2$

b. $\displaystyle\sum_{j=2}^{5} \frac{j-1}{j+1}$

c. $\displaystyle\sum_{k=0}^{6} \frac{1}{k!}$

Topic 22.1 Objective 6: Write a Series Using Summation Notation

Example 7:

Study the solutions for Example 7 parts a and b on page 22.1-20, and record the answers below.

Rewrite each series using summation notation. Use 1 as the lower limit of summation.

a. $2 + 4 + 6 + 8 + 10 + 12$

b. $1 + 2 + 6 + 24 + 120 + 720 + \ldots + 3{,}628{,}800$

Topic 22.2 Guided Notebook

Topic 22.2 Arithmetic Sequences and Series

Read the list of "THINGS TO KNOW" and review any concepts you are unfamiliar with.

Topic 22.2 Objective 1: Determine if a Sequence is Arithmetic

Write down the definition of an **Arithmetic Sequence** and provide one example.

In an arithmetic sequence the general term has the form $a_n = $ _____, where a_1 is _____ and d is _____.

Example 1:
Study the solutions for Example 1 parts a – d on page 22.2-4, and record the answers below. Watch the video for further explanation, if needed.

For each of the following sequences, determine if it is arithmetic. If the sequence is arithmetic, find the common difference.

a. $1, 4, 7, 10, 13, \ldots$ b. $b_n = n^2 - n$ c. $a_n = -2n + 7$ d. $a_1 = 14$, $a_n = 3 + a_{n-1}$

When the common difference of an arithmetic sequence is <u>positive</u>, the terms of the sequence _____ and the graph is represented by a set of ordered pairs that lie along a line with a _____ slope.

When the common difference of an arithmetic sequence is <u>negative</u>, the terms of the sequence _____ and the graph is represented by a set of ordered pairs that lie along a line with a _____ slope.

Topic 22.2

Topic 22.2 Objective 2: Find the General Term or a Specific Term of an Arithmetic Sequence

Rewrite the general term of an arithmetic sequence here: $a_n =$

Example 2:
Study the solutions for Example 2 parts a – c on page 22.2-7, and record the answers below. Watch the video for further explanation, if needed.

Find the general term of each arithmetic sequence, then find the indicated term of the sequence.

a. $11, 17, 23, 29, 35, \ldots$; a_{50}

b. $2, 0, -2, -4, -6, \ldots$; a_{90}

c. Find a_{31}

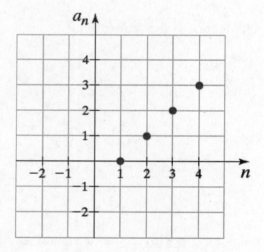

Topic 22.2

Example 3:
Study the solutions for Example 3 parts a and b on page 22.2-9, and record the answers below. Watch the video for further explanation, if needed.

Find the specific term of an arithmetic sequence:

a. Given an arithmetic sequence with $d = -4$ and $a_3 = 14$, find a_{50}.

b. Given an arithmetic sequence with $a_4 = 12$ and $a_{15} = -10$, find a_{41}.

Topic 22.2 Objective 3: Compute the *n*th Partial Sum of an Arithmetic Series

Write down the formula for the *n*th **Partial Sum of an Arithmetic Series.** Identify what each variable represents.

Topic 22.2

Example 4:
Study the solution for Example 4 part a on page 22.2-12, and record the answer below. Complete part b on your own. If your answer is incorrect watch the video to find your error

Find the sum of each arithmetic series.

a. $\sum_{i=1}^{20}(2i-11)$

b. $-5 + (-1) + 3 + 7 + \ldots + 39$

Topic 22.2 Objective 4: Solve Applications of Arithmetic Sequences and Series

Example 5:
Study the solution for Example 5 on page 22.2-14, and record the answer below.

A local newspaper has hired teenagers to go door-to-door to try to solicit new subscribers. The teenagers receive $2 for selling the first subscription. For each additional subscription sold, the newspaper will pay the teenagers 10 cents more than what was paid for the previous subscription. How much will the teenagers get paid for selling the 100th subscription? How much money will the teenagers earn by selling 100 subscriptions?

Example 6:
Complete Example 6 on page 22.2-15 on your own. Check your answer by watching the video.

A large multiplex movie house has many theaters. The smallest theater has only 12 rows. There are six seats in the first row. Each row has two seats more than the previous row. How many total seats are there in this theater?

Topic 22.3 Guided Notebook

Topic 22.3 Geometric Sequences and Series

Read the list of "THINGS TO KNOW" and review any concepts you are unfamiliar with.

Topic 22.3 Objective 1: Write the Terms of a Geometric Sequence

Write down the definition of a **Geometric Sequence** and provide one example.

In a geometric sequence the general term has the form $a_n = $ _____,
where a_1 is _____ and r is _____.

Make note of the differences between the graph of an arithmetic sequence and the graph of a geometric sequence (figure 3 on 22.3-5).

Example 1:
Study the solutions for Example 1 parts a and b on page 22.3-6, and record the answers below. Watch the video for further explanation, if needed.

a. Write the first five terms of the geometric sequence having a first term of 2 and a common ratio of 3.

b. Write the first five terms of the geometric sequence such that $a_1 = -4$ and $a_n = -5a_{n-1}$ for $n \geq 2$.

Topic 22.3 Objective 2: Determine If a Sequence Is Geometric

How can you determine if a given sequence is geometric?

Topic 22.3

Example 2:
Study the solutions for Example 2 parts a and b on page 22.3-8, and record the answers below. Complete part c on your own and check your answer by watching the video.

For each of the following sequences, determine if it is geometric. If the sequence is geometric, find the common ratio.

a. $2, 4, 6, 8, 10, \ldots$

b. $\dfrac{2}{3}, \dfrac{4}{9}, \dfrac{8}{27}, \dfrac{16}{81}, \dfrac{32}{243}, \ldots$

c. $12, -6, 3, -\dfrac{3}{2}, \dfrac{3}{4}, \ldots$

Topic 22.3 Objective 3: Find the General Term or a Specific Term of a Geometric Sequence

Rewrite the general term of a geometric sequence here: $a_n =$

Example 3:
Study the solutions for Example 3 parts a and b on page 22.3-10.

Example 4:
Study the solutions for Example 4 parts a and b on page 22.3-11, and record the answers below. Watch the video for further explanation, if needed.

a. Find the seventh term of the geometric sequence whose first term is 2 and whose common ratio is -3.

b. Given a geometric sequence such that $a_6 = 16$ and $a_9 = 2$, find a_{13}.

Topic 22.3 Objective 4: Compute the *n*th Partial Sum of a Geometric Series

Write down the formula for the ***n*th Partial Sum of a Geometric Series.** Identify what each variable represents.

Example 5:
Study the solution for Example 5 part a on page 22.3-15, and record the answer below. Complete part b on your own. If your answer is incorrect watch the video to find your error.

a. Find the sum of the series $\sum_{i=1}^{15} 5(-2)^{i-1}$

b. Find the 7th partial sum of the geometric series $8 + 6 + \frac{9}{2} + \frac{27}{8} + ...$

Topic 22.3 Objective 5: Determine if an Infinite Geometric Series Converges or Diverges

Study Table 1 on page 22.3-17, fill in the missing terms:

Looking at Table 1, it appears that as n increases, the value of S$_n$ _____.

Also notice that as n increases, the value of $r^n = \left(\frac{2}{3}\right)^n$ is getting closer to _____.

Write down the formula for the **Sum of an Infinite Geometric Series.** Identify what each variable represents. What must the value of r be restricted to?

If $|r| < 1$, then the infinite geometric series has a finite sum and is said to _____.

If $|r| \geq 1$, then the infinite geometric series does not have a finite sum and the series is said to _____.

Example 6:
Study the solutions for Example 6 on page 22.3-18.

Topic 22.3 Objective 6: Solve Applications of Geometric Sequences and Series

Example 7:
Study the solution for Example 7 on page 22.3-20, and record the answer below.

Suppose that you have agreed to work for Donald Trump on a particular job for 21 days. Mr. Trump gives you two choices of payment. You can be paid $100 for the first day and an addition $50 per day for each subsequent day. Or, you can choose to be paid 1 penny for the first day with your pay doubling each subsequent day. Which method of payment yields the most income?

Example 8:
Study the solution for Example 8 on page 22.3-21.

Amount of an Ordinary Annuity after the kth Payment

The total amount of an ordinary annuity after the kth payment is given by the formula

$$A = \frac{P((1+i)^k - 1)}{i}$$

where A = Total amount of annuity after k payments

  P = Deposit amount at the end of each payment period

  i = Interest rate per payment period

Example 10:
Complete Example 10 on page 22.3-27 on your own. Click on the video to check your work and answer.

Chie and Ben decided to save for their newborn son Jack's college education. They decided to invest $200 every 3 months in an investment earning 8% interest compounded quarterly. How much is this investment worth after 18 years?

Topic 22.4 Guided Notebook

Topic 22.4 The Binomial Theorem

Read the list of "THINGS TO KNOW" and review any concepts you are unfamiliar with.

Topic 22.4 Objective 1: <u>Expand Binomials Raised to a Power Using Pascal's Triangle</u>

Determine the expansion of $(a+b)^n$. Use the given values of n.

$n = 0$: $(a+b)^0 =$ 1

$n = 1$: $(a+b)^1 =$ 1a + 1b

$n = 2$: $(a+b)^2 =$ $1a^2 + 2ab + 1b^2$

$n = 3$: $(a+b)^3 =$

$n = 4$: $(a+b)^4 =$

$n = 5$: $(a+b)^5 =$

Fill in the missing blanks for Pascal's Triangle:

$n = 0$ 1

$n = 1$ 1   1

$n = 2$ 1   2   1

$n = 3$ 1   __   __   1

$n = 4$ 1   4   6   4   1

$n = 5$ 1   __   __   __   __   1

Continue the expansion for $n = 6$ and note the final answer in the expansion of $(a+b)^6$:

Topic 22.4

Example 1:
Study the solutions for Example 1 parts a and b on page 22.4-6, and record the answers below. Complete part c on your own and watch the video for further explanation, if needed.

Use Pascal's triangle to expand each binomial.

a. $(x+2)^4$ 	 b. $(x-3)^5$ 	 c. $(2x-3y)^3$

Topic 22.4 Objective 2: Evaluate Binomial Coefficients

Write down the **Formula for a Binomial Coefficient.**

Example 2:
Study the solutions for Example 2 parts a – c on page 22.4-10, and record the answers below.

Evaluate each of the following binomial coefficients.

a. $\binom{5}{3}$ 	 b. $\binom{4}{1}$ 	 c. $\binom{12}{8}$

How would you use a graphing calculator to evaluate the problems in Example 2?

Topic 22.4 Objective 3: Expand Binomials Raised to a Power Using the Binomial Theorem

Write down the **Binomial Theorem**.

Example 3:
Study the solutions for Example 3 parts a and b on page 22.4-11, and record the answers below. Watch the video for further explanation, if needed.

Use the Binomial Theorem to expand each binomial.

a. $(x-1)^8$

b. $(\sqrt{x}+y^2)^5$

Topic 22.4 Objective 4: Find a Particular Term or a Particular Coefficient of a Binomial Expansion

Write down the **Formula for the $(r+1)^{st}$ Term of a Binomial Expansion**.

Topic 22.4

Example 4:
Study the solution for Example 4 on page 22.4-14, and record the answer below. Watch the video for further explanation, if needed.

Find the third term of the expansion of $(2x-3)^{10}$.

Example 5:
Study the solution for Example 5 on page 22.4-15, and record the answer below. Watch the video for further explanation, if needed.

Find the coefficient of x^7 in the expansion of $(x+4)^{11}$.

Topic 23.1

Topic 23.1 Guided Notebook

Topic 23.1 Synthetic Division

Read the list of "THINGS TO KNOW" and review any concepts you are unfamiliar with.

Topic 23.1 Objective 1: Divide a Polynomial by a Binomial Using Synthetic Division

Review how to divide polynomials using long division from Topic 14.7.

If we divide a polynomial by a _____ in the form _____, then a shortcut method called _____ division can be used instead.

Watch the animation on page 23.1-3 to see how synthetic division compares to polynomial long division.

Example 1:
Study the solution for Example 1 on page 23.1-3, and record the answer below.

Divide $2x^4 + 9x^3 - 12x + 1$ by $x + 5$ using synthetic division.

Topic 23.1

Example 2:
Complete Example 2 on page 23.1-6 and record the answer below. Check your answer by clicking on the link. If your answer is incorrect, watch the video to find your error.

Divide $2x^4 - 3x^2 + 5x - 30$ by $x - 2$ using synthetic division.

Example 3:
Complete Example 3 on page 23.1-7 and record the answer below. Check your answer by clicking on the link. If your answer is incorrect, watch the video to find your error.

Divide $4x^3 - 8x^2 + 7x - 4$ by $x - \dfrac{1}{2}$ using synthetic division.

Read and summarize the CAUTION statement on page 23.1-7.

Topic 23.2 Guided Notebook

Topic 23.2 Solving Systems of Linear Equations Using Matrices

Read the list of "THINGS TO KNOW" and review any concepts you are unfamiliar with.

Topic 23.2 Objective 1: Write an Augmented Matrix

What is a matrix? Give an example.

The size of a matrix is determined by the number of _____ and _____.
A 2×3 (read *two by three*) matrix has _____ rows and _____ columns.

What is an augmented matrix?

Write the augmented matrix for the following system of equations:

$$\begin{cases} 3x + 2y = 11 \\ 8x - 9y = 4 \end{cases}$$

Example 1:
Study the solutions for Example 1 parts a and b on page 23.2-5 and record the answers below. Complete part c on your own and check your answer by clicking on the link. If your answer is incorrect, watch the video to find your error.

Write the corresponding augmented matrix for each system of equations.

a. $\begin{cases} -2x + 7y = 9 \\ 8x + 3y = 0 \end{cases}$

b. $\begin{cases} 6x - 3y = 11 \\ y = 2x - 5 \end{cases}$

c. $\begin{cases} 5x - 2y + z = 18 \\ x + 3z = -5 \\ 3y = 9 + z \end{cases}$

Topic 23.2

23.2 Objective 2: Solve Systems of Two Equations Using Matrices

When are two matrices **equivalent matrices**?

Record the **Row Operations**.
1.

2.

3.

Fill in the table below:

Notation	Meaning
$R_i \Leftrightarrow R_j$	
$kR_i \to \text{New } R_i$	
$kR_i + R_j \to \text{New } R_j$	

Watch the **popup** on page 23.2-8 to see an example of each row operation.

Describe a matrix in **row-echelon** form and give an example.

Example 2:
Study the solution for Example 2 part a on page 23.2-9 and record the answer below. Complete part b on your own and check your answer by clicking on the link. If your answer is incorrect, watch the video to find your error.

Use matrices to solve each of the following systems.

a. $\begin{cases} 2x - 5y = 26 \\ 3x + 2y = 1 \end{cases}$

b. $\begin{cases} y = 3x + 2 \\ 6x + \dfrac{1}{2}y = 6 \end{cases}$

Topic 23.2

Read and summarize the CAUTION statement on page 23.2-12.

23.2 Objective 3: Solve Systems of Three Equations Using Matrices

Record the steps for **Writing Row-Echelon Form (Three Rows)**

1.

2.

3.

4.

Example 4:
Study the solution for Example 4 part a on page 23.2-16 and record the answer below. Complete part b on your own and check your answer by clicking on the link. If your answer is incorrect, watch the video to find your error.

Use matrices to solve each of the following systems.

a. $\begin{cases} 2x + y - 3z = -3 \\ x + 2y + z = 4 \\ -3x + y - 4z = 1 \end{cases}$

b. $\begin{cases} 2x + 2y + z = 1 \\ 5x + 2y = 30 + 3z \\ 3x + 4z = -11 \end{cases}$

Topic 23.2

Example 5:

Study the solution for Example 5 on page 23.2-21, and record the answer below.

Use matrices to solve each of the following system.
$$\begin{cases} x+2y-3z=5 \\ -2x-3y+4z=-6 \\ 2x+4y-6z=10 \end{cases}$$

Read and summarize the CAUTION statement on page 23.2-24.

Topic 23.3 Guided Notebook

Topic 23.3 Determinants and Cramer's Rule

Read the list of "THINGS TO KNOW" and review any concepts you are unfamiliar with.

Topic 23.3 Objective 1: Evaluate a 2 x 2 Determinant

A **square matrix** has an equal number of _____ and _____.

Write the definition of the **Determinant of a** 2×2 **Matrix.**

Read and summarize the CAUTION statement on page 23.3-4.

Example 1:
Study the solutions for Example 1 parts a and b on page 23.3-4, and record the answers below.

Evaluate each determinant.

a. $\begin{vmatrix} 5 & 2 \\ 8 & 7 \end{vmatrix}$

b. $\begin{vmatrix} -9 & 3 \\ -4 & 2 \end{vmatrix}$

Topic 23.3 Objective 2: Use Cramer's Rule to Solve a System of Linear Equations in Two Variables

Record **Cramer's Rule for Solving Systems of Linear Equations in Two Variables**.

Topic 23.3

Watch the video on page 23.3-6 to see Cramer's Rule derived.

Example 2:
Study the solution for Example 2 part a on page 23.3-6 and record the answer below. Complete part b on your own and check your answer by clicking on the link. If your answer is incorrect, watch the video to find your error.

Use Cramer's Rule to solve each system.

a. $\begin{cases} 3x + 2y = 12 \\ 4x - 5y = -7 \end{cases}$
b. $\begin{cases} 6x + y = -2 \\ 9x - 2y = 11 \end{cases}$

If $D = 0$, the $x = \dfrac{D_x}{D}$ and $y = \dfrac{D_y}{D}$ are not _____, so Cramer's Rule does not apply. However, $D = 0$ tells us that the **system** is either _____ or _____.

Read through **Cramer's Rule with Inconsistent and Dependent Systems**.

Example 3:
Study the solution for Example 3 part a on page 23.3-9 and record the answer below. Complete part b on your own and check your answer by clicking on the link. If your answer is incorrect, watch the video to find your error.

Use Cramer's Rule to solve each system.

a. $\begin{cases} 2x - 5y = 7 \\ -4x + 10y = 11 \end{cases}$
b. $\begin{cases} 6x + 10y = -50 \\ -9x - 15y = 75 \end{cases}$

Topic 23.3 Objective 3: Evaluate a 3 x 3 Determinant

For every entry of a 3×3 determinant, there is a 2×2 determinant associated with it called its
_____.

For the following matrix, find the minor associated with b_1.

$$\begin{vmatrix} a_1 & b_1 & c_1 \\ a_2 & b_2 & c_2 \\ a_3 & b_3 & c_3 \end{vmatrix}$$

Record the definition for the **Determinant of a 3 × 3 Matrix**

Example 4:
Study the solution for example 4 on page 23.3-11.

Example 5:
Study the solution for Example 5 part a on page 23.3-14 and record the answer below. Complete parts b and c on your own. Watch the interactive video for the complete solutions.

Evaluate the determinant by expanding the minors for the given row or column.

$$\begin{vmatrix} 2 & 4 & -3 \\ 3 & 1 & 0 \\ -1 & -2 & 5 \end{vmatrix}$$

a. First column   b. Second row   c. Third column

Topic 23.3 Objective 4: Use Cramer's Rule to Solve a System of Linear Equations in Three Variables

Study **Cramer's Rule for Solving Systems of Linear Equations in Three Variables** as found on page 23.3-16.

Topic 23.3

Example 6:
Study the solution for Example 6 part a on page 23.3-17 and record the answer below. Complete part b on your own and check your answer by clicking on the link. If your answer is incorrect, watch the video to find your error.

Use Cramer's Rule to solve each system.

a. $\begin{cases} x - y + 4z = -5 \\ 6x - 27 + 5z = 1 \\ 3x + 2y + 2z = 10 \end{cases}$

b. $\begin{cases} x + y + z = 3 \\ 2x - 3y = 14 \\ 4y + 5z = -3 \end{cases}$

Study **Cramer's Rule with Inconsistent and Dependent Systems in Three Variables** as found on page 23.3-19.